MODERN
MOTORCYCLE
TECHNOLOGY

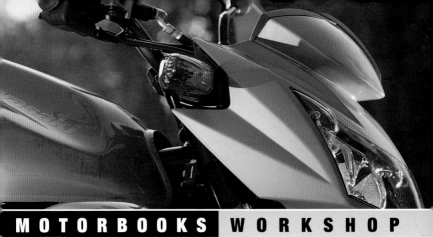

# MODERN
# MOTORCYCLE
## TECHNOLOGY

## HOW EVERY PART OF
## YOUR MOTORCYCLE WORKS

Massimo Clarke

*motorbooks*

ENGINES • PERFORMANCE • TRANSMISSION • CHASSIS
SUSPENSION • BRAKES • WHEELS AND TIRES

All pictures and diagrams in this book are from the archives of Massimo Clarke, unless otherwise indicated.

The publisher and author thank the following companies for their generous contributions: AE, Alfa Romeo, Aprilia, Benelli, BMW, Bosch, Brembo, Champion, Dell'Orto, Denso, Discacciati, Ducati, Exide, FIAMM, Harley-Davidson, Honda, Kawasaki, Keihin, KTM, Lancia, Mahle, Marzocchi, Michelin, Mikuni, Moto Guzzi, Moto Morini, Motul, MV Agusta, NGK, Pirelli, Suzuki, Toyota, and Yamaha.

Art Director: Giorgio Seppi
Editorial direction: Lidia Maurizi
Editorial coordination: Veronica Buzzano
Editing and pagemaking: Ervin s.r.l., Roma
English-language translation: Jay Hyams
English-language typesetting: William Schultz

First published in 2010 by **MBI Publishing Company** and Motorbooks, an imprint of MBI Publishing Company, 400 First Avenue North, Suite 300, Minneapolis, MN 55401 USA

Copyright © 2008 by Mondadori Electa SpA, Milano
Originally published in Italy as *Manuale Della Moto*

Motorbooks titles are also available at discounts in bulk quantity for industrial or sales-promotional use. For details write to Special Sales Manager at MBI Publishing Company, 400 First Avenue North, Suite 300, Minneapolis, MN 55401 USA.

To find out more about our books, visit us online at www.motorbooks.com

ISBN-13: 978-0-7603-3819-3

Printed and bound in Italy

# Contents

# Preface

Motorcycle technology has made great strides in recent years. New ideas have taken hold, while others that were already in widespread use have undergone a striking evolution. The constant rush toward higher and higher levels of performance has been accompanied by the need to address increasingly strict limitations in terms of noise and exhaust emissions. The developments that have taken place in the engine have been nothing less than astonishing, but they have been accompanied by equally important advancements in the components of the "cycle" part of motorcycles, beginning with the tires and suspension systems. Also of fundamental importance have been the adoption of fuel-injection systems and the large-scale application of electronics.

This manual was designed to serve the purpose of describing as clearly as possible every aspect of today's highly technological field of motorcycling. Its handy, compact format contains a high density of information. The structure and function of each of the many systems that compose a modern motorcycle are explained in clear language and from the point of view of an ordinary reader. A large section is dedicated to the technology behind each motorcycle part, the materials used in its creation, and the manufacturing methods employed. The text is accompanied by hundreds of color images and includes a series of sidebars that provide in-depth information on subjects of particular importance.

# The Motorcycle

Types
and Uses

# TYPES AND USES

A motorcycle can be divided into two parts, the "motor," meaning the engine and transmission, and the "cycle," composed of the frame, suspension, wheels, and brakes. In the past, all motorcycles were fundamentally quite similar despite the variety of uses for which they were made. Sport models were nothing more than faster versions of the basic model (using the same frame, suspension, and so on), perhaps with slight variations on the aesthetic level. The same applies even for models made for off-road use. Gradually, however, motorcycle models became more specialized, and today each motorcycle type sharply differs from the others in terms of appearance and also in terms of mechanics and features.

Naked (sometimes called "standard") motorcycles are without fairing, leaving the mechanics visible. Put to a wide variety of uses, they are as suitable for daily city use as for use on open roads, even trips. Indeed, they do not perform poorly if put to decidedly sporting uses. All told, they are versatile and multi-use to the point that they reach excellent speeds on race tracks. The frame is often of tubular construction and in some cases has a trellis structure. Standards usually have 17-inch wheels and a good braking system. The handlebars, whether low or medium in height, are almost always more or less straight.

Sportbikes are high-performance motorcycles typically with fairings.

■ The technical refinements of a modern high-performance motorcycle, such as this four-cylinder MV Agusta, are reflected in the complexity of the engine packaging.

■ Sportbikes like this GSX-R1000 put the rider in a racing position and are equipped with brakes and suspension systems worthy of a competition bike.

In recent years this category has come to include even higher performing bikes called superbikes, some of which can be considered street-legal race bikes, since only minimal modifications are necessary to ready them for racing. Sportbikes are designed for performance, not comfort, with the rider positioned low behind the windshield; the foot pegs are high and set back to the rear, and the clip-on handlebars are minimally adjustable. The "cycle" geometry (angle of the head tube, trail, and so on) put a premium on handling precision. Sportbike engines offer astonishingly specific power, and both the suspension and braking system are the best available on the motorcycle-parts market.

Superbikes are directly derived from the large-engine sportbikes. These are the champions on race courses all over the world.

An interesting segment of the motorcycle market is composed of machines designed for long-distance touring. These are made to accommodate two riders and offer cargo space along with notable comforts and excellent wind and weather protection.

■ The Kawasaki GTR 1400 is an excellent example of the sport tourer category of motorcycles, made for comfort and provided with accessories and bodywork to shield both driver and passenger.

■ Two of the major appeals of standard motorcycles are their versatility and practicality. This is a BMW R1200R.

■ For several years, large adventure-touring bikes like this powerful KTM 990 Adventure S have been popular with riders, to whom they offer excellent performance on both asphalt and off-road.

■ Numerous superbike competition machines are based directly on 1,000cc four-cylinder sportbikes, represented here by this Yamaha R1.

■ For cruisers, with their custom-style aesthetics, nothing is better than a large-displacement V-twin engine, often air cooled. Harley-Davidson is the king of this market.

■ With its modern technology on full display, this two-cylinder Morini is a beautiful example of a modern standard-style motorcycle.

These models are comfortable to ride and feature large fairings to provide protection for the rider, making them good in the rain. The handlebars are taller than on a standard machine and the large saddle is well shaped and padded; creature comforts are designed down to the smallest detail. In many cases the grips are electrically heated so they can be more comfortably used at low temperatures. Touring bikes are often large, with a generous wheelbase, so they tend to be heavy.

Enduro bikes, originally designed for use in off-road events, are all-terrain bikes that have long been popular for their great versatility and ease of handling. They are, in fact, suitable for a wide range of uses, from daily street riding to touring with two people and baggage to riding on dirt roads, sand,

or even on off-road routes, provided these are not too demanding. Enduros usually have wire wheels, narrow, slightly raised handlebars, and a comfortable, upright position for the rider. The suspension system offers lengthy travel, the front wheel is usually 19 or 21 inches, and the trail dimension is often long. Enduro engines typically have one cylinder (if the displacement is less than 650 cc) or two.

Cruiser bikes originated with certain American models of the 1950s and 1960s. Their characteristics include a longer, more raked fork, long trail, a stretched-out position for the rider with forward-mounted foot pegs, sometimes very high handlebars, and, in some cases, extended seats with a small backrest. The engines are usually V-twins that have relatively modest outputs compared to their displacement size, but offer high torque at low rpm and an unmistakable sound. Cruisers might have wire or cast wheels. The wheelbase is invariably long. In addition to these main road bike types, there are others, far more specialized, that are designed to serve specific "slices" of the market. They include supermotard bikes, competition off-road bikes, retro-style motorcycles, mini bikes, adventure touring bikes, choppers, and all those machines that are not road-legal, having been made specifically for roadracing and motocross.

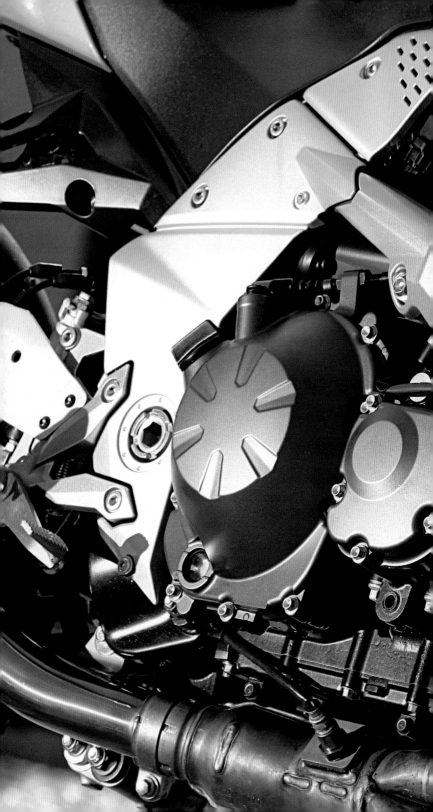

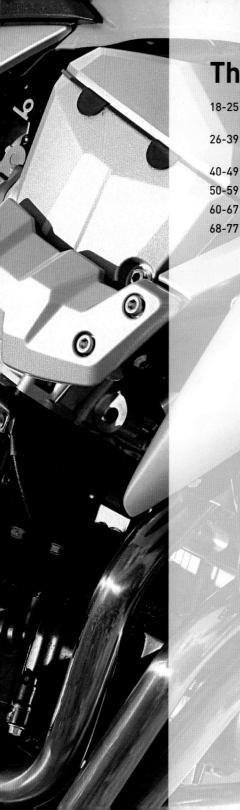

# The Engine

# STRUCTURE
# AND FUNCTION

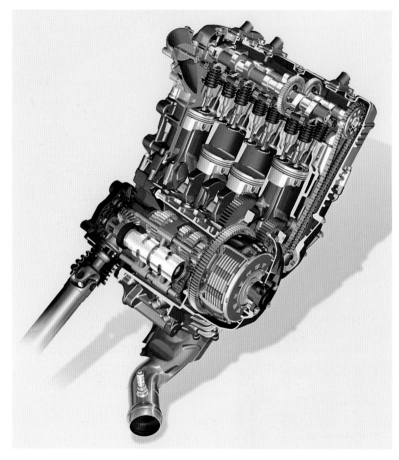

Modern motorcycles are most typically powered by four-stroke engines using spark ignition, also known as Otto-cycle engines after their inventor, Nikolaus August Otto. (The name also applies to two-stroke engines.) The fuel powering most bikes is gasoline, though a few countries use ethanol instead. Two- and three-wheeled vehicles powered by modest-performance diesel engines derived from industrial uses have been made in India, but they are mentioned here only as a curiosity.

■ The logical structure of BMW's modern high-performance inline four-cylinder engine is clearly visible in this illustration (BMW).

In giving a schematic description of a typical four-stroke motorcycle engine, we will make reference, for the sake of simplicity, to a vertical single-cylinder with only two valves. The structure is composed of both fixed and moving parts. First among the fixed parts is the crankcase, which houses the rotating

■ Single-cylinder engines are more compact and simpler from the construction point of view. This image highlights the arrangement and shapes of the components inside BMW's single (BMW).

## Basic Engine Terms

**BDC**—Acronym for bottom dead center, meaning the position at which the piston is located nearest the crankshaft. If the cylinder is vertical, it is the lowest position the piston reaches inside the cylinder. When it reaches BDC, the piston stops briefly before beginning to travel in the opposite direction, toward TDC.

**Bore**—The diameter of a cylinder opening, usually expressed in millimeters.

**Compression ratio**—When the piston rises to TDC, after having drawn the air/fuel mixture into the cylinder, it presses against the mixture, raising its pressure and temperature considerably. The amount of this compression depends on the ratio between the maximum volume of the piston available for the gases (piston at BDC) and the minimum volume (piston at TDC), as the gases become confined in the combustion chamber: This is the compression ratio.

**Displacement**—The volume displaced by a piston in its movement from one dead center to the other, expressed in cubic centimeters (less often in liters). The number is obtained by multiplying the area of a cross section of the cylinder by the stroke. It is thus easy to calculate once the bore and stroke of the engine are known.

**Engine displacement**—Also called "swept volume," this number is obtained by multiplying the displacement of a single cylinder by the number of cylinders in the engine.

**Stroke**—The distance, expressed in millimeters, between the two extreme ends, or dead points, reached by the piston in the course of its movement inside the cylinder.

**TDC**—Acronym for top dead center, meaning the position at which the piston is located farthest from the crankshaft. If the cylinder is vertical, it is the highest position the piston reaches inside the cylinder. When it reaches TDC, the piston stops briefly before beginning to travel in the opposite direction, toward BDC.

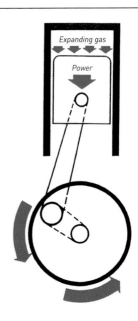

The piston moves up and down in the cylinder during the four strokes of its cycle. This side view shows the arrangement of the valves and ports in the cylinder head.

parts of the engine like the crankshaft as well as the bearings that support the crankshaft. There is then the cylinder, fixed above the crankshaft and mated to the head. Inside the head are the valves with their respective springs and the other parts involved in their movement. Also in the head are the combustion chamber and the openings for the intake and exhaust valves. The internal surface of the cylinder is the liner, and inside the piston moves up and down.

As noted above, located inside the fixed parts are the moving parts. These are the crankshaft with its connecting rod, the piston, and the parts of the valvetrain (valves, camshafts, rocker arms, or tappets). The big end of the connecting rod is connected to the crankpin of the crankshaft by way of a bearing, and the connecting rod small end is connected to the piston via a tubular bearing called the wristpin. The movements of the valves permit or prevent the passage of gases. The intake valve opens during the intake stroke admitting fresh fuel/air mixture, while the exhaust valve opens during the exhaust stroke, permitting the

In the four-stroke sequence, the power stroke is that of expansion, during which the gases drive the piston down. The piston is connected to the crankshaft by the connecting rod (Honda).

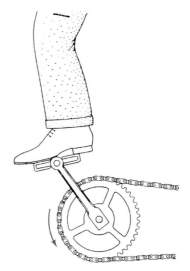

The connecting rod operates on the crankpin the way a leg acts on a bicycle pedal; the up-and-down motion of the rider's knees is transformed into the spinning movement of the bicycle's sprocket (Honda).

expulsion of burnt gases from the cylinder. During the compression and combustion strokes, both valves are closed.

The structure of two-stroke engines is similar, but the head is reduced to a sort of simple cover since it does not house any moving parts and does not "host" the intake and exhaust valves. Instead, the walls of the cylinder have a series of openings, called ports, for intake, exhaust, and transfer.

**The Four-Stroke Cycle**

The engine performs a sequence composed of four strokes, each stroke being the action of the piston traveling the full length of its cylinder. The four strokes are intake, compression, combustion, and exhaust. In the case of a four-stroke engine, these strokes are performed in two revolutions of the crankshaft, or in 720-degrees rotation. The sequence takes place as follows:

*Intake:* The down-stroke of the piston, moving from top dead center (TDC) to bottom dead center (BDC), creates a vacuum in the cylinder that draws air/fuel mixture into the cylinder through the intake tube (the valve of which has opened).

*Compression:* On the up-stroke of the piston, moving from BDC to TDC, the air/fuel mixture is compressed. At the end of this stroke the pressure and temperature of the air/fuel mixture (trapped in the combustion chamber)

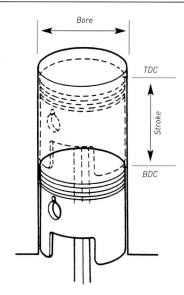

■ Bore is the diameter of the cylinder, and stroke is the distance between the two end points reached by the piston as it moves inside the cylinder (Suzuki).

reaches values on the order of more than 20 bar of pressure and 380 to 500 degrees Celsius.

*Combustion (power stroke):* At the end of the compression stroke, as the piston nears TDC, an electric spark flashes between the electrodes of the spark plug and ignites the air/fuel mixture, thus beginning combustion. This causes the development of a tremendous amount of heat and also a strong rise in

■ The sequence of the four-stroke engine. From left: intake, compression, combustion, exhaust.

■ This drawing presents the shape and arrangement of the various parts that compose a four-cylinder inline engine with twin camshafts. Also visible is the primary transmission with gears, the multidisk clutch, and the gearbox (Kawasaki).

the temperature and pressure in the cylinder. In a contemporary, high-performance four-stroke engine the pressure can reach past 85 bar. During the combustion stroke the piston is forced down to BDC by the pressure of the gas. This is the only power stroke of the four-stroke cycle and now is when mechanical energy is supplied to the crankshaft.

*Exhaust:* After reaching BDC at the end of the combustion stroke, the piston again rises to TDC, expelling the burnt gases from the cylinder, for which the exhaust valve opens, permitting release of the gases.

### The Two-Stroke Cycle
Today, four-stroke engines dominate the field of motorcycles. Two-stroke engines are largely confined to mopeds

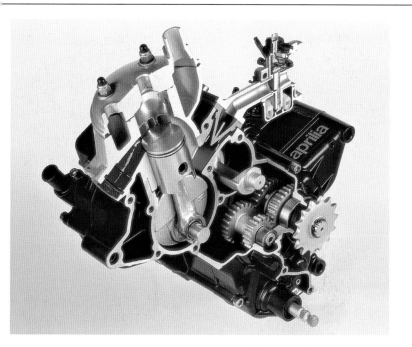

and small-engine scooters along with a few competition machines (Grand Prix motorcycles of 125 and 250cc and various off-road bikes).

The advantages of two-stroke engines include their mechanical simplicity, their compactness and limited weight in relation to displacement, and the enormous specific power they can deliver. The drawbacks are high fuel consumption in relation to power delivered and exhaust emissions so high that these engines are rendered

■ The structure of two-stroke engines, which today have only a modest and limited use, is very different from that of their four-stroke cousins (Aprilia).

unacceptable unless adapted to a direct-injection system. They also tend to have shorter life spans than four-stroke engines.

In two-stroke engines, intake does not take place in the cylinder, but rather in the crankcase in which the crankshaft

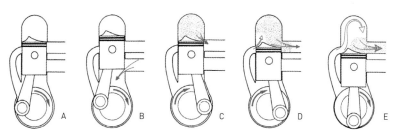

■ The four phases of the sequence of two-stroke engines take place in a single revolution of the crankshaft. Intake takes place in the crankcase. A = compression, B = intake in the crankcase, C = beginning of exhaust (at the end of expansion), D = beginning of transfer, E = transfer in full development.

## Simple and Compact

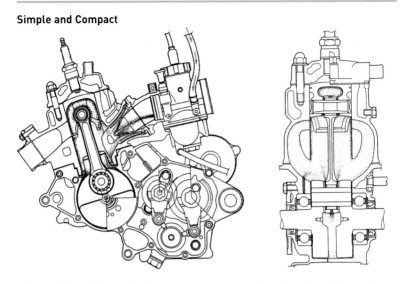

These two views make possible a clear understanding of the mechanical simplicity of two-stroke engines. Since they have a power stroke every turn (rather than every two turns), they can deliver far more specific power than four-stroke engines (Honda).

revolves. After being drawn into this chamber, the air/fuel mixture is sent to the upper part of the cylinder through the transfer port. The engine can thus be thought of as being divided in two parts: the part beneath the piston acts as pump, while the four strokes of the cycle take place in the upper part. Unlike what happens in four-stroke engines, in this case the strokes all take place in a single turn of the crankshaft, meaning 360 degrees of rotation. The cycle takes place as follows:

*The piston rises from BDC toward TDC.* As the piston rises, a low-pressure area is created in the crankcase that draws in a fresh air/fuel mixture (fresh air from the outside mixed with fuel inside the carburetor). In the upper part of the cylinder, the rising piston shuts off the transfer port and immediately after that the exhaust port (located a little higher) closes. The rising piston progressively compresses the air/fuel mixture that earlier entered through the transfer port.

*The piston descends from TDC to BDC.* Combustion takes place above the piston, creating pressure that forces the piston downward toward BDC. At a certain point as it moves downward the upper edge of the piston uncovers the exhaust port, and the burnt gases begin to flow out of the cylinder.

At the same time, the descent of the piston reduces the space available in the crankcase for the fresh gases, resulting in a kind of "precompression." Shortly after this, having moved farther toward BDC, the piston uncovers the transfer port, and the air/fuel mixture rises from the crankcase into the cylinder, where it takes the place of the burnt gases—and to a certain degree contributes to scavenging those burnt gases, for which reason the fresh gases are said to perform a kind of "washing" of the cylinder.

### Reed Valves and Transfer Ports
In two-stroke engines, the intake port, which connects the carburetor to the

crankcase, must have a valve (or other system) able to permit or prevent the passage of the fresh air/fuel mixture at the right time in the cycle. If it did not have such a valve, the crankcase could not function as a pressurization chamber for the air/fuel mixture.
An automatic, one-way valve called a reed valve is used in most two-stroke engines. In some competition models a rotary disk valve is used instead. The number and arrangement of the transfer ports are very important in terms of engine performance. There are usually five of these (four larger ones on the sides, "of delivery," plus one of "correction" located opposite the exhaust port).

Two-stroke engines were formerly crossflow-scavenged, with transfer and exhaust ports on opposite sides of the cylinder and a deflector atop the piston to direct the intake and discharge. This system has been replaced by loop-scavenging in which transfer ports direct the flow of the gases to the

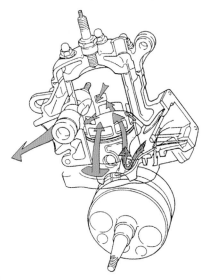

■ The arrows indicate the movement of gases during the transfer phase in the cylinder of a modern two-stroke engine (Yamaha).

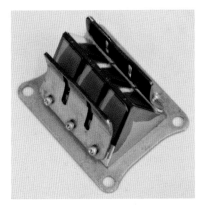

■ In almost all of today's two-stroke engines the intake is controlled by a reed valve that functions automatically.

combustion chamber, from where they are deflected downward. Even so, the scavenging is anything but perfect because of the inevitable loss of a considerable amount of the fresh mixture, which leaves the engine with the burnt gases (thus depriving the engine of fuel while also polluting the atmosphere). At the same time, a notable quantity of burnt gases is not expelled and remains inside the cylinder. In high-performance two-stroke engines, great importance is given to fitting them with exhaust valve systems or expansion chambers, which make it possible to best exploit the waves of pressure so as to fill the cylinder with fresh air/fuel mixture.

# PERFORMANCE AND OUTPUT

## Performance and Output

Engines using the four-stroke Otto cycle (as well as diesel engines) are energy transformers. They function by converting some of the heat produced by combustion into mechanical energy. Following the flash of spark from the spark plug's electrodes, a flame is propagated that crosses the combustion chamber, igniting the mass of air/fuel mixture contained there. Combustion is an oxidation process that takes place with numerous intermediate passages and is notably exothermic, meaning it is

■ These are the moving parts of a BMW parallel two-cylinder engine with twin overhead camshafts and four valves per cylinder (BMW).

accompanied by a considerable production of thermal energy.

The temperature in the combustion chamber rises rapidly (during combustion values on the order of 2,500 degrees Celsius are reached), and the pressure also increases.

This pressure acts on the piston, driving it forcefully toward BDC. (The combustion stroke is the power stroke.) It is important to emphasize that the increase in pressure is extremely rapid, given that combustion takes place in a short time but also in a gradual manner.

In synthesis, what is provided to the engine by the fuel is chemical energy (potential) that combustion transforms into thermal energy (heat), part of which is converted into mechanical energy and can thus be used to drive the crankshaft assembly.

**Torque**

Torque is a force that produces torsion and thus involves an axis of rotation. More precisely, torque is a force that acts on a body from a distance (the distance between the body's axis of rotation and the point of application of the force itself). Tightening or loosening a lid (like that of a jar) involves the application of torque. Having failed to

■ The crankshaft and connecting rod transform the reciprocating up-and-down movement of the pistons into rotary motion (BMW).

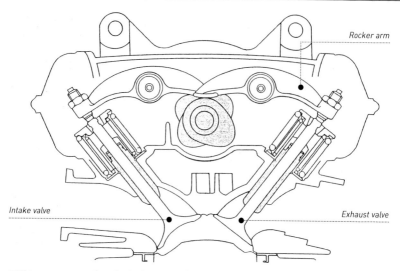

Rocker arm

Intake valve

Exhaust valve

■ This cross section of a cylinder head shows the arrangement of the camshaft and the two valves. The intake valve controls the flow of fresh air/fuel mixture that enters the cylinder, and the exhaust valve controls the burnt gases that exit it (Yamaha).

## Basic Elements

**Energy**—Energy is usually defined as the ability to perform work, in the physical sense. It appears in various forms (such as thermal, mechanical, and electrical) that can be converted into one another. Movement energy is most often kinetic energy. It is expressed in joules: 1 J = 1 N x 1 m. (The newton, indicated by the N, is a unit for measuring force.)

**Pressure**—Pressure is a force applied over an area. The unit of measurement for pressure is the pascal (Pa), but given the values at play in the field of mechanics, the unit most often used is the bar. 1 Pa = 1 N/m2. 1 bar = 100,000 Pa.

**Work**—Work is the force used to move an object. It thus has the same physical dimensions as energy. If there has been no movement, no work has been done. Aside from the work that results in movement, there is also "resistant" work, which brakes movement. The piston performs passive work during the phases of intake, compression, and exhaust; it receives energy from the gases when these expand, following combustion. Work is also measured in joules.

■ The crown of the piston receives the pressure of gases during the combustion stroke. A "clean" configuration of this component is always advantageous.

loosen a large nut with a wrench, you may well succeed if you insert the wrench handle in a long tube, creating an extension. Applying the same amount of force, but from a greater distance from the axis of rotation, creates a greater amount of torque. In the metric system, torque is measured in units of Nm (newton-meters). It is important to point out that torque can be applied

■ The larger the passageways, the better the "breathing" of the engine at high rpm. Large valves that move a considerable distance off their seats are thus an advantage in trying to obtain elevated performance.

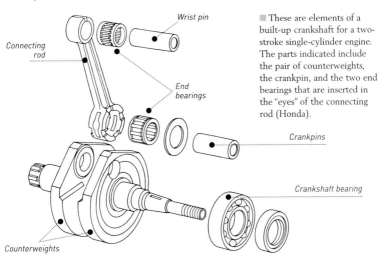

Wrist pin

Connecting rod

End bearings

Crankpins

Crankshaft bearing

Counterweights

■ These are elements of a built-up crankshaft for a two-stroke single-cylinder engine. The parts indicated include the pair of counterweights, the crankpin, and the two end bearings that are inserted in the "eyes" of the connecting rod (Honda).

without any movement taking place (as when the large nut would not move prior to use of the extension; despite the lack of movement, force from the wrench was being applied).

Torque can be increased or diminished through the use of gears, chains, or belts. A cyclist encountering a hill will change to a lower gear; in this way, applying the same force to the pedals, a greater amount of torque will be sent to the rear wheel. This is accomplished, however, at the price of a loss in speed; the rate at which the cyclist pedals remains the same, but the bicycle moves forward more slowly. What does not vary (with the exception of a small loss due to inevitable friction) is the product of the torque for the speed of rotation. Diminishing the speed of rotation by way of a lower gear increases the torque, and vice versa.

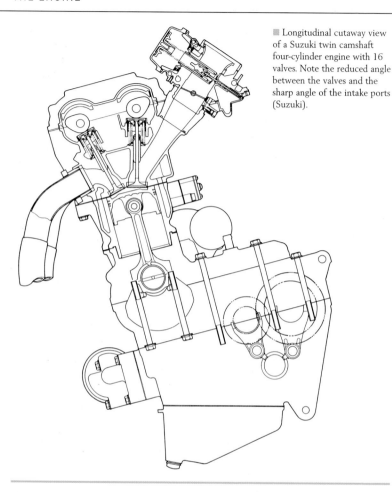

Longitudinal cutaway view of a Suzuki twin camshaft four-cylinder engine with 16 valves. Note the reduced angle between the valves and the sharp angle of the intake ports (Suzuki).

## From two to four valves per cylinder

Here is a comparison between a hemispherical combustion chamber with a two-valve cylinder head and a roof-shaped head with four valves. Note that in the four-valve head the spark plug is centrally located.

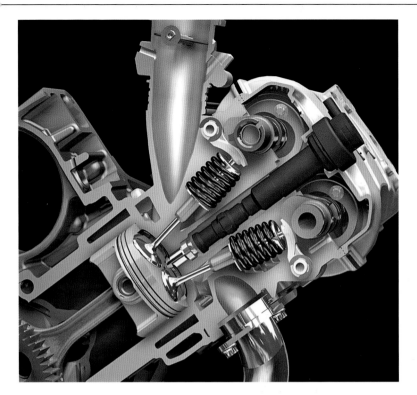

## Power Output and Flywheel Mass

The pressure that acts on the crown of the piston multiplied by the surface area of the piston, assumed to be perfectly smooth (meaning the area of the transverse section of the cylinder), equals the force with which the gases drive the piston toward BDC during the combustion stroke. This force is transmitted to the crankshaft by the wrist pin and the connecting rod. It does not act along the axis of the shaft but instead on the crankpin. The distance between the point of application of the force and the axis of the crankshaft's rotation is called the crank radius; it corresponds to half of the stroke. Thus, in addition to "collecting" the force and transforming it into torque, that is, into the force of torsion, the connecting rod/crankshaft unit converts the reciprocating up-and-down movement of the pistons into the rotating movement of the crankshaft, and vice versa. In fact, during the

■ It is in the cylinder head that power is "created." This side view of a high-performance engine shows the cams and the finger-type rocker arms, the impressive size of the intake port, and the compactness of the combustion chamber (BMW).

passive strokes of the operational sequence (intake, compression, exhaust), it is the crankshaft that moves the pistons.

The delivery of torque is pulsating and what the crankshaft receives is the average quantity produced by every four-stroke cycle at the various rpm. The irregularity of torque delivery diminishes as the number of cylinders in the engine increases. The irregularity is thus greatest in single-cylinder engines, which require large flywheel masses that have the function of absorbing the energy of movement during the active phase (when the speed

31

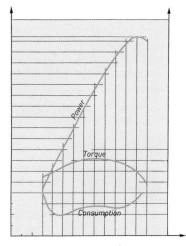

*Engine rotations in rpm*

The characteristic curves of an engine show the progression of torque and power on the basis of engine rotation speed. Also included here is the level of consumption. Such information is collected on an engine test stand with the throttle fully open.

of rotation of the crankshaft tends to increase) and then delivering it during the passive phases (when the speed tends to diminish). Thanks to the presence of an adequate flywheel mass (incorporated in the crankshaft or applied to it in the form of a discoidal element of the appropriate size), the function of the engine becomes more "round" and regular at low speeds.

## Torque and Power

The pressure that pushes against the crown of the piston during the combustion stroke is not constant but

varies, rapidly diminishing as the piston approaches BDC. (The highest value is reached shortly after TDC.) For this reason it is customary to use the average value for reference. The higher this number, the more "vigorous" are the power strokes that take place inside the cylinder and as a consequence the greater is the amount of torque supplied by the engine. The pressure of the gases (and therefore also the torque) is tied to the "breathing" of the engine, meaning the quantity of air/fuel mixture taken into the cylinder during each intake stroke. As will be seen later, since the quantity of this breathing (meaning its volumetric output) changes with variations in the engine's rpm speed, the same happens with torque output. The progress of torque on the basis of engine rpm can be measured and shown as a torque curve; it can also be experimentally assessed on an engine test stand.

The average amount of pressure acting on the piston during the combustion stroke is also related to the

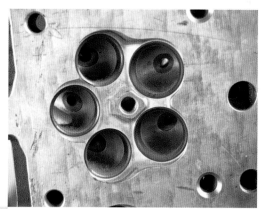

■ Heads with five valves per cylinder (three intake and two exhaust) have combustion chambers with a less favorable configuration than do those with four valves.

efficiency of the engine as a transformer of energy. It is related to how efficiently the engine uses the energy provided by the combustion of the air/fuel mixture (in other words, its thermal efficiency).

Power is the rate at which something is (or can be) accomplished. In terms of a motorcycle engine, power means the mechanical energy that the engine is able to supply within a given unit of time. Since the kilowatt, a multiple of watts, is the "official" unit of measure according to the International System of Units (SI), power should be expressed in kW, but in reality preference is still often given to horsepower

(hp): 1 hp = 0.746 kW.

Although very different physical quantities, torque and power are closely related. The power of an engine is nothing other than its torque multiplied by its speed of rotation. It thus does not vary along its "route" from the engine to the rear drive wheel, no matter what the gear ratio might be.

**Characteristic Curves**

Like torque, power varies with engine rotation speed, and like torque there is a power curve to track its value at all rotation speeds. The torque curve rises progressively with the increase in

## Engine Parameter

**Fractioning**—Putting aside total engine capacity and the stroke/bore relationship, a greater number of cylinders makes it possible to achieve more revolutions per minute, which aside from greater mechanical stress means greater power. This is also favored by the fact that the total surface of the pistons is greater. On the other hand, fractional engines are structurally more complex, and the size and weight of the engine increase as well.
**Specific power**—To determine how much an engine is "charged" and to compare engines of different displacements, reference is often made to specific power, which is usually expressed in horsepower per liter. To determine specific power, divide engine power output by engine capacity, expressed in liters. The four-stroke engines used in superbikes reach specific powers on the order of 200 horsepower per liter (at speeds of more than 13,000 rpm). MotoGP bikes have faster rotation speeds and can achieve 270 horsepower per liter. The two-stroke engines of the 125 and the 250 Grand Prix bikes go as high as 400 horsepower per liter.
**Bore/stroke relationship**—Engines are referred to as being "square" when the diameter of the cylinder bore is equal to the length of the stroke. They are called short stroke when the bore is greater than the stroke and long stroke (or "under square") when the opposite is true. With the exception of various "old-style" or

custom engines (such as classic Harley-Davidsons with their pushrod and rocker-arm valvetrains), motorcycle engines are short stroke. This is because reducing the bore and stroke relationship makes it possible to install larger valves and to reach faster rotation speeds using the same mechanical stress. In other words, in order for the engines to turn faster and to have large-diameter gas lines (indispensable conditions for reaching high levels of specific power) it is necessary for the relationship to be low. It is interesting to note that the situation is very different with two-stroke engines, and the best results, in terms of performance, are obtained with nearly square bore and stroke relationships.
**Mean piston speed**—In a single engine revolution the piston moves both up and down the cylinder, thus making two strokes. Knowing the rotation speed makes it easy to determine the speed at which the piston is moving. This speed can be an important reference since it is a somewhat reliable indication of the mechanical stress to which the components of the crankshaft are subjected. In modern high-performance engines, values on the order to 21 to 22 meters per second are common. Competition engines can exceed 25 meters per second, while for the most part the pistons of standard motorcycles and touring bikes do not exceed more than 19 to 20 meters per second.

■ Torque and power are measured on an engine test stand, which is also used to tune the ignition and the fuel injection (the "mapping" of the bike's central processing unit).

■ A flow bench is used in head design to optimize engine breathing, to achieve a highly elevated volumetric output. During design the testing is done not to an actual head but to a replica made of resin.

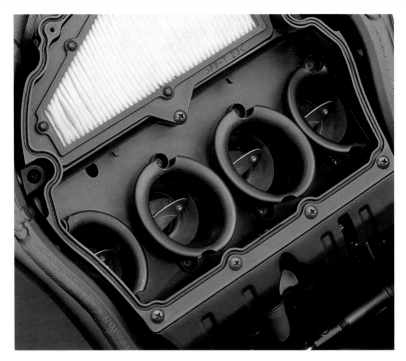

rotation speed until it reaches its highest value, after which it descends with greater or lesser gradualness. The power curve continues to rise, having reached its highest value, even after torque has begun to diminish. This happens because the increase in engine rotation speed more than makes up for the diminution of torque. In other words, the product of torque times rotation speed continues to increase even if less rapidly than before. At a certain engine rotation speed, power reaches its highest value. If the rotation speed is further increased, power decreases (in which case the engine is said to be operating "over revved"). The curve descends more or less rapidly. A further increase in engine rotation speed is not able to compensate for the diminution of torque, which becomes increasingly conspicuous since the engine is entering a true "respiratory" crisis.

The shapes of the curves of torque and power (known as the "characteristic curves" of the engine) are of great

■ The throttles on fuel-injected engines are the butterfly type.

interest since they reveal the "character" of the engine's output. These curves reveal, for example, if (and to what degree) the engine is "responsive" and requires an intense use of gears or if it is flexible, with a great "draw" even at low engine speeds and thus with a broad field of use.

### Engine Efficiency

Engine efficiency is a simple concept, dealing as it does with the relationship between what is actually obtained against what might have been obtained ideally; put another way, it is the difference between the power that is delivered and what has been furnished to the engine.

There are basically three types of engine efficiency: thermal, mechanical, and volumetric.

35

■ This cross section of a cylinder head and cylinder shows the state of the art in the field of high-performance engines. An essential feature is large intake ports, which divide in two inside the head, following a more or less straight course (Suzuki).

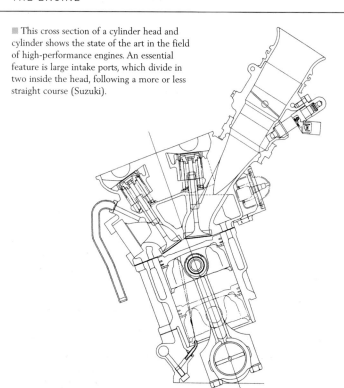

## Knocking

Since heat output (and with it engine performance) increases with an increase in the compression ratio, the compression ratio is taken to very high levels in supercharged engines. This path can only be taken only so far, however. When the compression ratio goes beyond a determined value-limit (which varies from engine to engine), it inevitably leads to knocking. This can have serious consequences for the engine. Knocking is the sudden, almost explosive combustion of part of the air/fuel mixture not yet reached by the flame front ignited by the spark plug. Consequently, an important aspect of fuel is its antiknocking power, expressed by the octane rating. The higher the octane rating, the higher the compression ratio that can be used without risk of knocking.

Thermal efficiency is itself composed of several other efficiencies (of combustion, availability, cycle), but this is not the place to go into such details. It can be defined as the relationship between the mechanical energy "swept" by the pistons and the energy theoretically available, meaning the energy "contained" in the fuel fed to the engine and then transformed into heat by combustion. Thermal efficiency tells us the efficiency with which the engine uses the energy that is made available to it (in other words, how capable it is as a "transformer" of energy). This efficiency increases, even if not linearly, with the compression ratio. Among the parameters that influence it are the configuration of the combustion chamber and the sizes of the cylinders.

Not all of the mechanical energy "swept" by the pistons reaches the transmission. Part is lost through inevitable friction and another part through

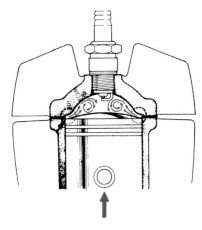

Adequate turbulence in the combustion chamber is advantageous because it speeds combustion; the turbulence, called "squish," is composed of many small vortexes generated by the expulsion of gases from certain areas when the piston nears TDC (Honda).

Volumetric efficiency is an expression of how well the engine "breathes." It is composed of the relationship between the actual amount of air that enters the cylinders with every intake stroke and the maximum amount of air that theoretically could enter. In other words, volumetric efficiency is the relationship between the quantity of air aspirated and the ideal quantity, taking into consideration ambient pressure that would occupy a volume equal to the engine's displacement. As is clear, as volumetric efficiency increases, the "energy" of the individual power strokes that take place

pumping (this being the passive work performed by the pistons to intake the gases and then expel them from the cylinders).

Friction occurs between moving parts and those fixed (as well as between parts in reciprocal movement, one against another), such as the piston/piston ring group and the cylinder liner, the crankpins on the crankshaft and the bearings on the crankshaft and the connecting rod, and so on. Further energy is lost through the operation of the valvetrain and to run the oil and water pumps, and another quantity of energy is lost through oil "sloshing" and internal ventilation.

The relationship between the mechanical energy that "exits" the engine and is delivered as power to the transmission and the power that is developed in the pistons by the expansion of gases constitutes an engine's mechanical efficiency. This efficiency diminishes with increases in the speed of engine rotation. For this reason, in high-performance engines, which reach very high speeds of rotation, sliding surfaces are reduced to a minimum and special measures are taken to limit, as far as possible, mechanical losses.

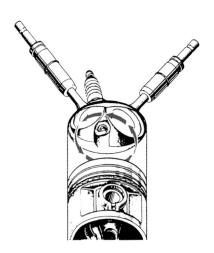

"Swirl" is well aimed intake turbulence that can be obtained with the right shape and arrangement of the intake port (Suzuki).

## Combustion and Knocking

The four diagrams in the left column give a schematic view of the process of normal combustion. The diagrams on the right present combustion with knocking: before the flame front has reached it, a part of the fresh air/fuel mixture suddenly ignites (General Motors).

inside the engine grows and thus also the power that, at that given rotation speed, the engine provides. The output changes with variations in engine speed rotation, reaching its highest value at a given rotation speed (that coincides with that at which there is maximum torque), after which it drops in a more or less gradual way.

The shape of the volumetric efficiency curve is determined by the timing of the valvetrain and by the

characteristics of the intake and exhaust systems, meaning the configuration and size of the exhaust and ports. One might think that, as with other outputs, the maximum value theoretically obtainable could be established, but that of volumetric efficiency is a case apart. Opportunely exploiting both the energy of the gas columns that travel inside the intake and exhaust systems as well as the waves of pressure, it is in fact possible to reach values slightly greater than 100 percent, in correspondence to certain

■ Modern high-performance engines are devices of great technological refinement in which all the components have been designed to obtain the best overall performance (Benelli).

rotation speeds. In Otto-cycle engines, power output is controlled by acting on the volumetric efficiency by means of the throttle; breathing is perfectly free (and thus the efficiency is highest) only when the throttle is fully open, which happens when the throttle is opened all the way.

# ENGINE DESIGN

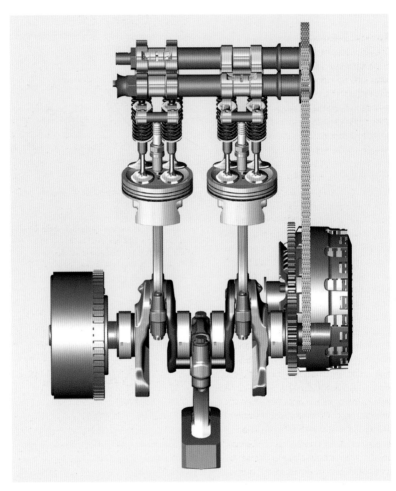

The simplest engines, not surprisingly, are those with only one cylinder. Light and compact in relation to their engine capacity, they cost less to manufacture than multi-cylinder engines since they involve fewer parts and less labor. They dominate the world of small engines as well as the enduro and motocross motorcycle classes (which must be agile and speedy and are thus fitted with engines whose weight and size have

■ The vibrations generated by parallel-twin engines can be eliminated only by using dynamic balancers. Visible here in a central position is the balancer used by BMW on its 800cc engines.

been greatly reduced). Single-cylinder engines are also used by most scooters.

The crankshaft has a single crankpin to which is mounted a single connecting rod, and the crankshaft is typically the built-up type (which proves convenient from both the construction and economic points of view). The crankcase will also be simple, with only two easy-to-fit roller bearings to support the crankshaft.

## Dynamic Balancers

The Achilles' heel for single-cylinder engines is the fact that they cannot be balanced by way of simple counterweights on the crankshaft. The parts that cannot be balanced are those in alternating motion (the pistons, piston pins, and some of the connecting rod). If on the side opposite the crankpin the crankshaft is fitted with an additional mass equal to that of the moving parts, the counterweight is perfect only when the piston is located at the two dead points. When the crankpin is located at a 90-degree angle to these there is a strong imbalance (enough to produce unacceptable vibrations) since the crankshaft comes to have, from the opposite side, an unbalanced mass that "pulls" outward by way of centrifugal force.

The opposite happens if the crankshaft is not provided with any counterweight mass (to offset the masses in alternating movement) on the side opposite the crankpin. In that case, perfect balance occurs only when the pin itself is at 90 degrees in relation to the dead points, with total imbalance when the pin corresponds to the dead points. As a result, a compromise is made with an "intermediate" solution, fitting the crankshaft with counterweight masses that balance, in general, 55 to 60 percent of the mass in alternating movement. It is thus inevitable that notable vibrations will occur when the engine is running. That does not pose special problems if the engine is small or is destined for sport uses. If, however, it is of larger displacement and is made for road use, the situation changes. The vibrations become unacceptable and must be adequately

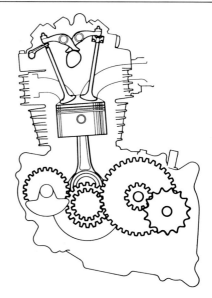

This is a cross section of a single-cylinder large-displacement engine. The auxiliary balance shaft is located ahead of the crankshaft, which operates it by way of a pair of meshed gears.

Extra balance shafts are usually fitted with an eccentric mass that generates unbalancing forces moving in the opposite direction to those generated by the crankshaft, which are thus offset.

offset. This is done through the use of a dynamic balancing component, usually a balance shaft, which is an eccentric, weighted shaft that is operated by the crankshaft by way of gears or a toothed belt and that turns in the opposite direction of the crankshaft. Such balance shafts can drastically diminish vibrations, and even if they do not eliminate them completely, they prove in general more than adequate to the needs.

■ This drawing of a parallel twin engine with a crankshaft with crankpins at 180 degrees illustrates the shape and arrangement of the counter-rotation balance shaft with two eccentrics (Kawasaki).

## Parallel Twin Engines

The architecture of two-cylinder engines can vary. Parallel twin engines, those in which the cylinders are side by side and have parallel axes, permit the adoption of two different solutions in terms of the position of the crankpins on the crankshaft. They can be aligned (crankpins at 360 degrees), in which case the two pistons move together toward the same dead points (in other words, rise and fall together, inside their cylinders), or they can be positioned in such a way that while one piston moves toward TDC the other descends toward BDC (the crankpins

■ BMW's parallel-twin features a crankshaft fitted with crankpins at 360 degrees. A dynamic balancer composed of an extra "balancing" connecting rod, turned downward and attached to an oscillating mass helps quell vibration.

at 180 degrees). With the crankpins arranged at 360 degrees, the distance between the power strokes becomes perfectly equal (the "detonations" follow one another every 360 degrees of rotation of the crankcase), creating a particularly "round" and pleasing engine sound. In terms of vibrations, however, such an engine behaves like a single cylinder, meaning it cannot be balanced by way of a counterweight on the crankshaft. In this case dynamic balancers are used to lower the level of vibration. The system used calls for the adoption of one or two extra shafts fitted with eccentric weights. BMW's 800cc twin offers a unique and effective engine balancer of a different design. In this case, the crankshaft rests on four support bearings; at its center is an eccentric to which is attached an extra "balancing" connecting rod pointing down (in the opposite direction of the cylinders), the end of which is hinged to a balancing rocker. This arrangement generates forces of inertia similar in magnitude but in the opposite direction to those produced by the engine's two connecting rod-piston groups. In this way there are no notable vibrations. The engines of some Yamaha scooters employ a similar system called a reciprocating balancer, although in this case instead of an oscillating mass at the end of an extra connecting rod there is a "false" piston.

In parallel twins with crankpins at 180 degrees, one piston rises to TDC while the other descends to BDC. One might think this would make it possible to achieve excellent balance, but in reality the layout cannot be called

perfect. The axes of the two cylinders are inevitably at a certain distance one from the other, and this generates unbalanced torque that is the source of vibrations that can have such intensity they become decidedly unpleasant. To offset such vibration, a counter-rotation balance shaft is used, fitted with two eccentric masses. In engines with this architecture the power strokes are not uniformly distanced but occur at every 180 degrees—540 degrees—180 degrees of rotation of the crankshaft. When the engine is running at its slowest or at low rotation speeds, this irregularity is markedly noticeable from the sound of the exhaust: The engine is not round and instead tends to "gallop."

### V-twins and Boxer Twins

For the past several years the preferred architecture for use on two-cylinder engines has been the V. The angles actually used by engineers differ according to the model in question, and today those angles fall between a

■ A cross section of a 60-degree V-twin Rotax engine of 1,000cc: This sophisticated balance system calls for two main balance shafts (the one in front of the crankshaft is clearly visible) plus an additional shaft, of modest size, located in the rear head (Aprilia).

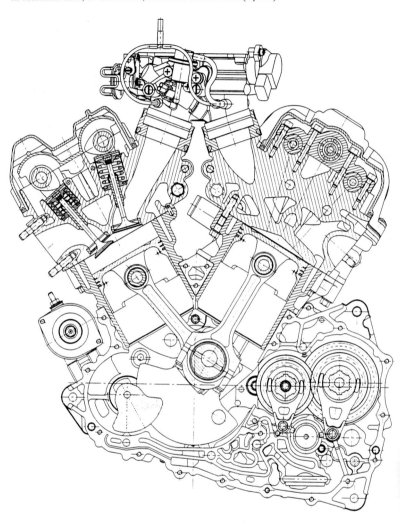

maximum of 90 degrees and a minimum of 45 degrees. The most advantageous solution in terms of balance and thus the control of vibrations is that of 90 degrees. Both Ducati and Moto Guzzi have adopted that angle for their cylinders. (Moto Guzzi's transverse V-twins are unique among motorcycles, while Ducati, in keeping with the classical school, uses a longitudinal V, meaning the axis of rotation of the

crankshaft is transverse to the frame.) Moto Morini produces a very modern engine in which the V between the cylinders is 87 degrees, which thus fits this category. (The slight deviation from 90 degrees was a packaging adjustment.) Suzuki and Honda have also made large twins using this architecture.

Two-cylinder 90-degree V engines are characterized by fairly good balance (the forces of inertia of the first order

■ Good balance is obtained in this Honda 52-degree V-twin through the use of a crankshaft with two offset crankpins.

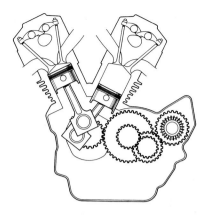

are perfectly balanced), although it is not perfect. Furthermore, if the V is longitudinal, the engine tends to have a considerable length. On the upside, its width may well be reduced markedly, though this is true for all V-twins. The crankshaft is short and rigid, and the two connecting rods are mounted side by side on the same crankpin. The power strokes are not uniformly distanced but follow every 450 degrees— 270 degrees—450 degrees. Diminishing the angle between the cylinders improves the longitudinal compactness of the engine but degrades the balance.

■ In this BMW boxer twin (below) the forces of inertia of both the first and second order are balanced. The two cylinders are not coaxial, however, which generates perceptible vibrations. BMW eliminates them with a balance shaft.

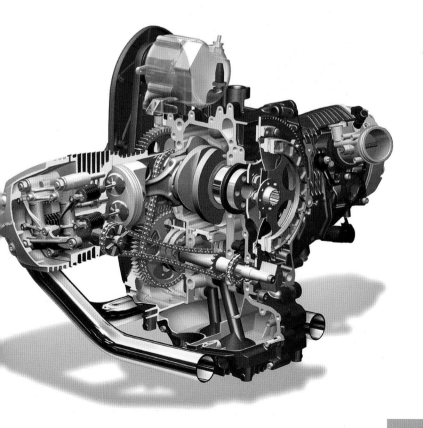

Two methods are employed to reduce the vibrations in V-twins. The first is the "classical" solution and calls for the use of an auxiliary crankshaft (less often two). The second adopts a crankshaft fitted with not one but two staggered crankpins; as is obvious, this presents advantages in terms of construction simplicity and containment of costs, but it is less suitable for use on high-performance engines since the crankshaft is less rigid and strong, given the diameter of the pins. (The crankpins are separated by a narrow unsupported central flywheel.) Neither of these solutions achieves an equal distance between the power strokes.

Boxer twins form a category of their own. BMW is the best known proponent of this arrangement, which features a longitudinal axis of the crankshaft and horizontally opposed cylinders that protrude from either side of the crankcase. With this arrangement, the balance is nearly perfect, with very low vibration levels and equidistant engine power strokes. This arrangement is also well suited to air cooling, since the heads and the cylinders are fully exposed to the airstream. The crankshaft rests on two supports and is fitted with two crankpins arranged at 180 degrees. The side-to-side bulk is considerable, while the length and vertical heights are well contained. This arrangement requires an automobile-type transmission, with separate gears and shaft-driven final drives.

**Three and Four Cylinders**
Motorcycles featuring three-cylinder engines have been manufactured several times over the history of the motorcycle, and while not built in great

■ Three-cylinder engines are inherently well balanced. In contemporary versions, however, a balance shaft is employed to further dampen vibration and to obtain engine operation completely free of vibration (BMW).

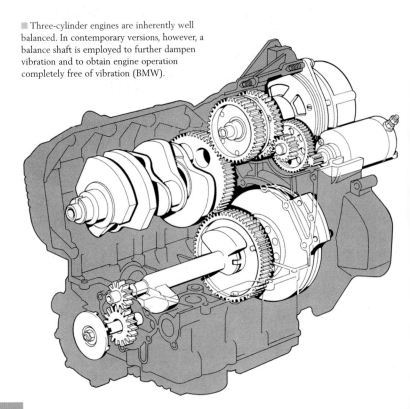

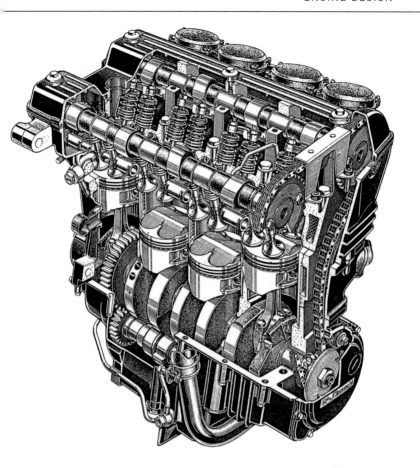

■ Some Japanese large-displacement, inline four-cylinder engines are fitted with a small extra balance shaft that turns in the opposite direction of the crankshaft (Kawasaki).

numbers, these instances are none-theless important to note. There have been several interesting examples of V-configuration engines featuring three cylinders (two parallel side cylinders and a third slightly inclined with respect to them). This arrangement is no longer in use today, but excellent inline triples are made by Triumph and Benelli. In these motorcycles the crankshaft, which rests on four main bearings, has the crankpins arranged at 120 degrees, making possible perfect equidistancing

between the power strokes and providing a good degree of balance.

In fact, with such a configuration the forces of inertia resulting from the alternating masses in movement are perfectly balanced. The same is not true of the torque, which if the displacement is considerable can cause somewhat bothersome vibrations, although they are not particularly intense. To offset this vibration an auxiliary balance shaft that turns at the same speed as the crankshaft but in the opposite direction is fitted. (It is usually located in the front part of the crankcase.)

Inline, four-cylinder engines offer numerous advantages, thus making them one of the most common motorcycle engine configurations particularly above a certain displacement. They operate smoothly, they are practically vibration-free, and they make it possible to obtain much higher levels of power than can be reached by two- or three-cylinder engines with the same displacement. The power strokes are uniformly spaced and follow every 180 degrees (meaning the output of torque is less pulsating than with three-cylinder engines and, most of all, than with twins). The crankshaft rests on five main bearings and has its crankpins arranged at 180 degrees: When the engine is running, the two central pistons move toward TDC while the two pistons on the sides move toward BDC, and vice versa. The balance, as good as it is, however, is not perfect. The primary forces of inertia are balanced, as is the torque, but the secondary forces of inertia are not. These secondary vibrations, however, are relatively more mild, and with most four-cylinder engines there is no need to use dynamic balancers of any kind.

■ For its inline-four-cylinder used in the K 1200 (R and S) series, BMW engineered a highly refined balance system that uses twin counter-rotation balance shafts, visible in this image below the crankshaft.

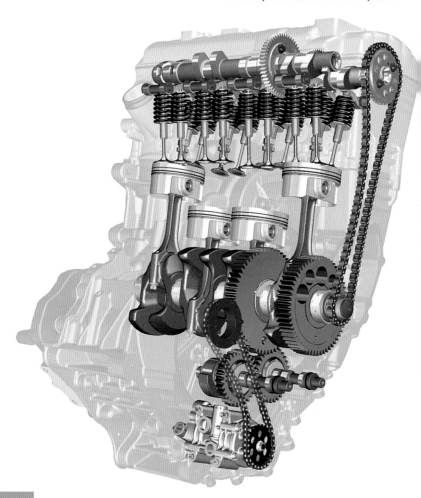

■ The four-cylinder V architecture is of notable interest and is used by various MotoGP bikes, as well as on a pair of very modern superbikes made by Aprilia and Ducati (Ducati).

Even so, the use of a small counter-rotation balance shaft can be a good idea particularly for engines with large displacements. Such shafts are usually positioned in the front part of the crankcase. Recent four-cylinder BMWs and some Kawasaki motorcycles are fitted with two balance shafts that make it possible to achieve absolutely perfect balance.

In addition to the configurations described, there are others that employ four cylinders in a V or opposing layout, and even five or six cylinders, but such arrangements are used only rarely.

# VALVETRAIN

In four-stroke engines, the valvetrain refers to the mechanical parts, such as the valves themselves, that control the flow of the gases that enter and leave the cylinder during the intake and exhaust strokes.

Motorcycle engine valves are opened by camshafts. The movement of the valves is controlled by the shape of the camshaft's lobes, which push against intermediate valvetrain parts that serve to absorb the lateral force generated by the lobes' rotation. If these intermediate parts rotate on a shaft and move up and down, they are rocker arms. If they are moved by metal tubes, they are pushrods. Valves are closed by a spring, the notable exception being Ducati motorcycle engines, which use the desmodromic valve mechanism (see page 57).

In most modern high-performance engines one or two camshafts are housed in each head, and the rocker arms (or pushrods) are in direct contact with the valves. In this way the mass in reciprocating motion is reduced to a minimum, and the

■ This is a head for a four-cylinder high-performance engine: The chain operates only the exhaust-valve camshaft; the intake camshaft is driven by a gear on its shaft, which mates to a corresponding gear on the exhaust camshaft (BMW).

system has the highest level of rigidity.

### Advancing and Retarding Valve Movement

The valves do not suddenly jump from the closed position to that of maximum lift in correspondence to the two dead-center points. (Doing so would not be mechanically feasible, since it would result in unacceptable levels of stress.) Instead they rise gradually and use a certain amount of time to move from the closed position to that of maximum lift, and then of course the same happens as the valves move back from maximum lift to the closed position. Put another way, valves cannot be subject to overly elevated acceleration because it would involve unacceptable forces of inertia. Thus, in order to have

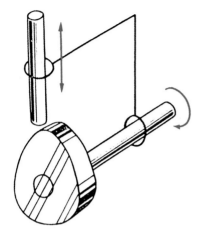

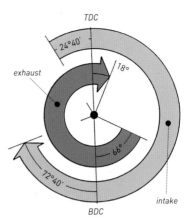

■ This drawing illustrates the way the camshaft lobe transforms rotating movement into the linear movement of the rocker arm (Honda).

■ A typical valve timing diagram for a four-stroke engine: In this case, the exhaust has a duration of 264 degrees and an intake of 277 degrees 20 minutes (Alfa Romeo).

the intake valve already rising from its seat when the force exercised by the piston becomes considerable, in the first part of the stroke, from TDC to BDC, it must begin to open before the piston reaches TDC at the end of the exhaust stroke. The same happens with its closing, which must end considerably later than BDC at the end of the intake stroke. The situation is absolutely the same for the exhaust valve. Thus, the beginning of the opening of the valves

and the end of their closing do not coincide with the dead points, which means the phases of intake and exhaust are not confined to their relative strokes. Even the compression stroke begins with a certain delay with regard to BDC, meaning precisely when the intake valve comes in contact with its seat, and thus it does not coincide with the compression stroke but has a smaller angular duration. The same holds true for the combustion phase

■ Many engines have only a single overhead camshaft, almost always chain-driven. In the example here, the valves are activated by two-arm rocker arms (Aprilia).

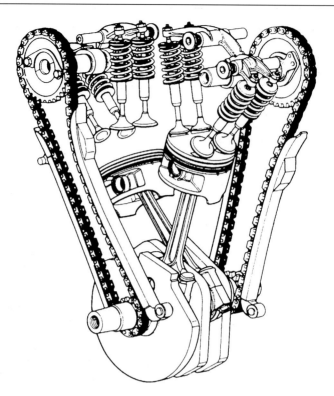

■ This single-camshaft V-twin has two silent chains, one for each head and thus for each camshaft (Suzuki).

and stroke. (The exhaust valve begins to open before BDC is reached.) This situation is very advantageous in terms of performance since it makes it possible to best exploit the energy of the gas column that enters and exits the cylinder and the waves of pressure that travel inside the ports.

**Inertia of Gases and Pressure Waves**
The intake valve begins to move from its seat well before the piston has arrived at TDC at the end of the exhaust stroke. The exhaust valve finishes closing after the piston has moved past TDC. This means that for a certain period, during TDC at the end of the exhaust stroke, both valves are partially open. This is known as valve overlap. This is advantageous in terms of scavenging the combustion chamber: The drawing

action exercised by the exhaust contributes to putting in movement the fresh air/fuel mixture, which then enters the cylinder and "sweeps" the chamber, freeing it of burnt gases. The intake phase begins even before the piston has begun to descend from TDC to BDC. Obviously this functions best only when the engine is running at certain speeds.

The intake valve does not fully close until a good while after the piston has passed BDC. Because of their inertia, the fresh gases continue to enter the cylinder even after the piston has changed the direction of its movement and has begun to rise toward TDC. It is clear that, in terms of refilling the cylinder, it is best if the valve finishes closing precisely when the gas column has stopped arriving. Also in this case, a certain delay in closing provides the best results in terms of performance, at a certain speed of engine rotation. The exhaust valve begins to open considerably before BDC. This is advantageous since the burnt gases begin to pour into the exhaust port thanks to their pressure, exiting in great quantity even before the piston has reached BDC; as a consequence, when the piston rises toward TDC, it has less work to do to expel the remaining exhaust gases. The advanced opening of the exhaust valve causes a reduction in the effective length of the combustion phase, but it must be pointed out that in the last part of the combustion stroke the gases have already done most of their work on the crown of the piston; the pressure in the cylinder is by then somewhat low and furthermore the angle between the connecting rod and the crank web on the crankshaft is disadvantageous.

Thus, very little efficiency is lost. Aside from the inertia of the gas column, modern high-performance engines fully exploit the pulsating phenomena that take place inside the intake and exhaust systems. Clearly, if a strong positive wave arrives just when the intake valve is nearly finished closing, the refilling improves. (It is as though a true "fluid piston" pushed more of the mixture into the cylinder.)

During the exhaust phase, it is important for a wave of negative pressure to reach the valve during the period of valve overlap, just when it is about to close. As for exploiting pressure waves to increase the volumetric output of the engine, it is clear that, with a certain amount of anticipation in the opening of valves and delay in their closing, the best results (for every given configuration of the intake and exhaust systems) will occur at a certain speed of engine rotation.

### Timing and Performance

Advancing and retarding the opening and closing of the valves in relation to the dead points of the cylinder constitutes valvetrain timing, which is usually expressed in degrees of crankshaft rotation. Timing that involves greater advancing and retarding makes it possible to obtain improved performance at high engine rpm; such timing is typical for sportbike engines that have very high specific power but not a strong power range at low and medium speeds. On the contrary, modest advancing and retarding are suitable for lower performance engines that do not rotate at high rpm and that provide modest performance in relation to their displacement (but that offer power over a broad rpm range). Valve timing can be expressed graphically with a circular or spiral

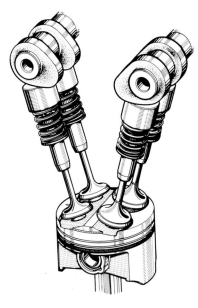

■ By far the most common arrangement in high-performance engines is to have bucket tappets (the cylindrical lifters) between the camshaft lobes and the four valves of each cylinder (Suzuki).

■ In the most recent sport version of its boxer twins, BMW has a double camshaft with conical lobes, finger-type rocker arms, and valves in a slightly radial arrangement.

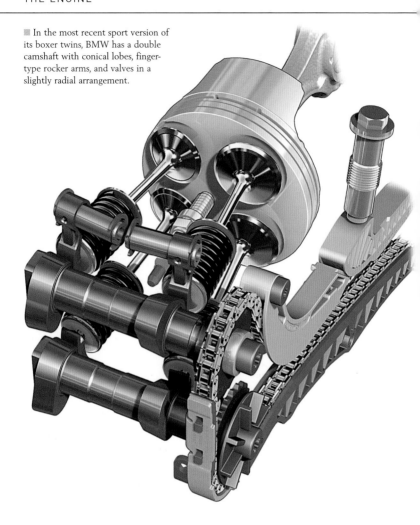

diagram in which the duration of the various phases is expressed visually. (Such graphs are called valve-timing diagrams.)

**How Many Valves?**

Since the earliest days of motorcycling, engines have been designed with two valves for each cylinder, one intake and one exhaust. In the 1960s, Honda produced competition models with four valves (two intake and two exhaust). This arrangement, though certainly not pioneered by Honda, met with great success from a technical point of view

and was soon adopted by other manufacturers for their Grand Prix motorcycles. Over time, four valves per cylinder was adopted by all manufacturers for their racing machines as well as for their higher performance standard models.

The use of four valves makes it possible to effectively make larger diameter ports for the passage of gases, leading to improved intake at high engine rpm and with less loss from pumping. Furthermore, the individual valve-pushrod (or rocker arm) units are lighter, which makes it possible for the

## Key Words

**Dual overhead cam**—Valvetrain configuration, indicated by the abbreviation DOHC, with two camshafts per head. Between the lobes and the valves are simply bucket tappets or rocker arms.

**Lift**—Lift is the distance a valve opens, meaning the distance between the head of the valve and its seat when open. The valve lift curve graphically indicates the amount of this movement in terms of the angle of rotation of the crankshaft. (It indicates the rocker arm ratio.) From this it is possible to calculate both the speeds and the accelerations (positive and negative) to which the valve is subject, on the basis of the rotation of the crankshaft. The maximum lift of the valve coincides with that of the cam when between the two pieces there is only a lifter. If rocker arms are being used, the relationship between the length of the arms must be taken into account.

**Pushrods and rocker arms**—Valvetrain configuration in which the camshaft is in the crankcase and operates the valves by way of lifters, or tappets, which control pushrods and rocker arms (the latter above the cylinder heads) that actuate the valves. Known variously as a pushrod engine or an overhead-valve engine, it is indicated by the abbreviation OHV. This configuration in unsuitable for high-performance engines that must reach high rpm. Today it is used only in some custom twins, "old-style" motorcycles, or those bikes based on a "classic" design.

**Single overhead cam**—Valvetrain configuration, indicated by the abbreviation SOHC, with a single camshaft per head. The valves are actuated by the lobes by way of a rocker arm with two arms. The mass in reciprocating motion is greater than in a dual cam configuration, and there is less rigidity in the operational system.

For several years Honda has been making large single-cylinders with single-camshaft valvetrains and four radial valves. (Note the use of two rocker arms for each of them.)

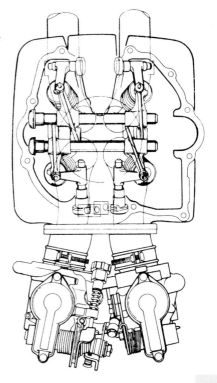

engine to reach higher speeds of rotation. Put simply, by using four valves per cylinder instead of two, greater specific power can be obtained.

### Drive Systems

The camshaft rotates at half the speed of the crankshaft. The crankshaft can drive the camshaft in various ways. The most common system used to drive the camshaft employs a chain, typically a roller chain or a Morse-type chain.

If the camshaft is overhead (above the valves), the cam chain will be

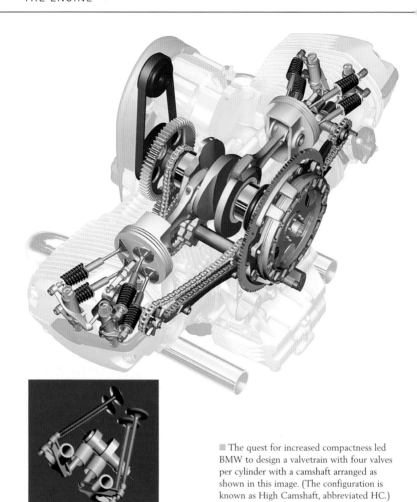

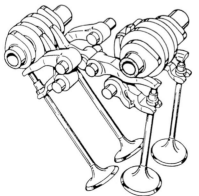

■ The quest for increased compactness led BMW to design a valvetrain with four valves per cylinder with a camshaft arranged as shown in this image. (The configuration is known as High Camshaft, abbreviated HC.)

■ The desmodromic arrangement used by Ducati, shown here in a twin-camshaft version, uses two rocker arms and two cams for each valve.

## Desmodromic Valvetrain

Sometimes valves are not closed by springs but by a mechanical system that acts in a manner analogous to the system that opens valves but in the opposite sense. Such is the desmodromic valvetrain used by Ducati in both race models and production models, with some variations but always functioning on the same basic principle. For every valve there are two cams with "doubled" lobes: One actuates the opening (acting on a conventionally shaped and located rocker arm), the other the closing. This closing rocker arm has a special shape and is positioned below the other rocker and upside down. Its forked tip acts on a small plate that is part of the valve and pulls it back after it has reached its maximum lift.

relatively long. Because the chain follows an irregular path, its path should incorporate an automatic tensioner and antivibration guides. The cam-drive system can involve more than one stage. For example, in the four-cylinder MV Agusta engine there is a first stage with a reduction gear entrusted to a pair of meshed gears and then a second stage that uses a chain. In some cases, in order to allow a more compact head in a twin-camshaft engine, cam-drive systems are used that call for a gear for every camshaft meshed with a third gear, central and located a little lower, that is operated by the chain. To take a further example, in BMW's boxer twins, a first chain drives an auxiliary shaft that runs two chains that transfer motion to the camshafts located in two heads.

Unlike automobile engines, toothed belts are rarely used in motorcycles.

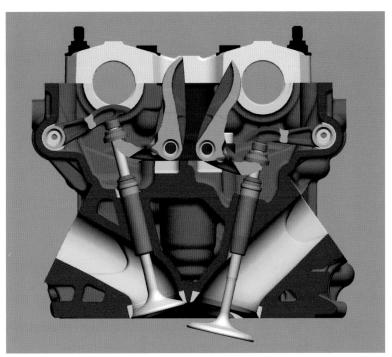

■ The further evolution of the Ducati head has led to the arrangement above, with the opening rocker arms (finger-type) turned and a notably diminished angle between the valves.

■ By far the most commonly used method of driving a dual camshaft layout uses a chain that operates directly on both camshafts (Kawasaki).

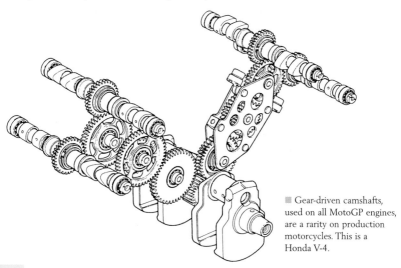

■ Gear-driven camshafts, used on all MotoGP engines, are a rarity on production motorcycles. This is a Honda V-4.

■ In certain cases, to reduce oscillation of the chain, a more expensive two-stage system is used, which involves a pair of shorter chains (Kawasaki).

■ The toothed belt, so widespread in automobiles, finds little use on motorcycle engines. An exception is the Ducati twin.

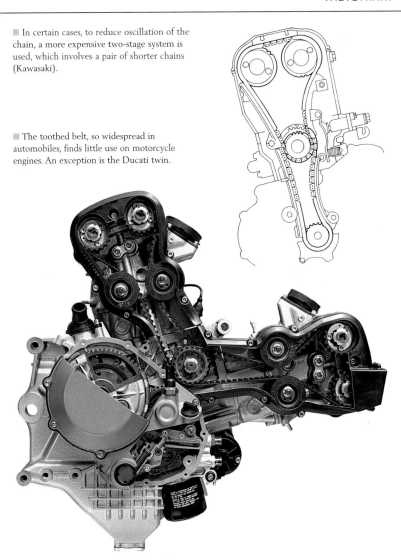

An exception is the two-cylinder Ducatis, which are fitted with a belt for each cylinder. (The belts are driven by an auxiliary shaft located in the crankcase.) The camshafts in competition engines are often driven by cascading gears; this solution, costly and sophisticated, is rarely applied to production engines.

# COOLING SYSTEMS

The engine is an energy converter. It transforms part of the heat created in its cylinders into mechanical energy, which is sent to the rear drive wheel first by means of the motorcycle's primary drive, which transfers engine power from the crankshaft to the clutch and gearbox, and then by means of the final drive. Unfortunately, as a "transformer" the engine is not overly efficient: only a little more than 30 percent of the available heat is actually used in a four-stroke Otto engine, even less in a two-stroke engine. The remaining roughly two-thirds of the energy theoretically available is lost, in part through exhaust and in part in the form of heat absorbed by the engine's metallic walls. In fact, after the engine has been running only a short time its walls would reach unacceptably high temperatures if the engine was not equipped with an adequate system to cool and dissipate the heat.

■ A Kawasaki motocross bike with two vertical-flow radiators in aluminum alloy

During combustion the gases inside the engine reach temperatures that exceed 2,000 degrees C. These gases are in contact with the inner walls of the head and the cylinders (as well as the valve heads and the crowns of the pistons), and a great amount of heat is imparted to these parts during the combustion and exhaust strokes, as well as, in part, during the compression stroke. This heat moves through the walls and is then dissipated by the exterior of the engine parts involved (air cooled) or by the fluid that runs through special openings (water cooled) inside the engine. It is essential that the engine not exceed certain temperatures because in doing so the mechanical character of various materials can undergo unacceptable

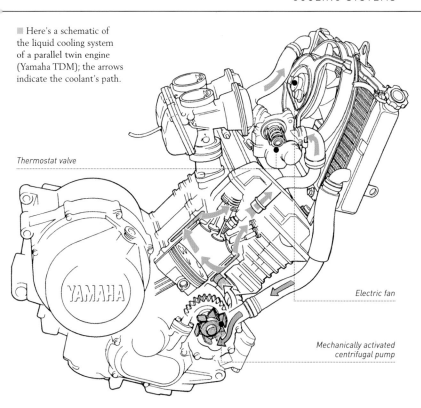

■ Here's a schematic of the liquid cooling system of a parallel twin engine (Yamaha TDM); the arrows indicate the coolant's path.

*Thermostat valve*

*Electric fan*

*Mechanically activated centrifugal pump*

deterioration. Excessive heat would cause a reduction in the clearance space between moving parts, and it could also cause serious distortion of components. Furthermore, the oil would no longer be able to adequately perform its function because of an accentuated loss of viscosity or a serious decline in its lubricating qualities.

## Fins to dissipate heat

Head of an air-cooled two-stroke engine fitted with a dense row of parallel fins

This picture of a two-stroke engine cylinder clearly shows the width of the fins.

The water passages for engine cooling are clear in this cylinder cross-section from a two-stroke engine.

### Cooling Systems

Engines can be cooled by air or by water. Some prefer to say "liquid" instead of "water" since the fluid running through a motorcycle engine is in fact a mixture of distilled water and antifreeze along with anticorrosive additives (or instead of these a pre-mixed coolant). In both systems it is always air that truly "carries away" the heat, whether this is done directly or indirectly. (In water-cooled engines the water serves only to transport the heat from the engine to radiators, where it is dissipated by air.) In addition to the quantity of heat dissipated by the motorcycle's cooling system, further heat is dissipated by cooling fins and some is carried off by the recirculating oil. Indeed, some engines are constructed in such a way that the cooling function of the oil assumes great importance; such cooling systems are referred to as "mixed" (air/oil). In such cases, aside from the usual fins, there is usually an oil radiator of considerable size.

Whichever system is used, the transfer of heat from high-temperature gases to the metal walls and past those walls by the cooling fluid happens by forced convection and is related to the differences in temperature, the extension of the surfaces, and the heat transfer coefficient. Heat is transmitted through the walls by conduction. When the surfaces are in contact with water, the heat transfer coefficient is much higher than when the cooling is done by air. For this reason, in order to achieve an adequate dissipation of heat, the external walls of the head and the cylinder in air-cooled engines are covered with a series of cooling fins of notable size. In this manner the surface area touched by the coolant, in this case air, is greatly increased, on average by about 15 to 20 times.

The shape, number, and size of the cooling fins are critical to engine cooling and must be designed with care. The speed at which the motorcycle is moving forward contributes to the passage through the fins of a copious amount of air. In air-cooled scooter engines with this cooling system, the head and cylinder are out of the direct air stream thus a fan is used to create a flow of air that is directed against the engine components by special ducting.

The heads of four-stroke engines are highly stressed by heat. Clearly visible here are the wide openings for the passage of coolant.

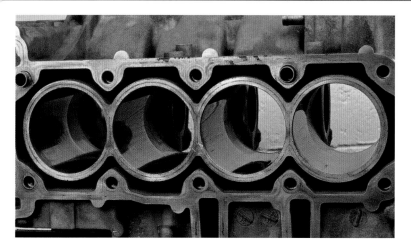

■ This is an open-deck cylinder block of the Siamese type (the adjacent cylinder bores form a unitary structure) surrounded by passages for coolant.

## Water Cooling

All engines with particularly high power are water cooled. Water cooling provides a more vigorous dissipation of heat than does air cooling, and it also makes it easier to control the temperatures of the various engine parts, keeping them within defined limits. Water cooling also facilitates a more even temperature throughout the engine, a factor that minimizes the risk of distortion. Water cooling also offers

■ In many high-performance motorcycles the radiator is a cross-flow and has a curved shape.

## Temperature Control

After the warm-up phase, which follows ignition and takes several minutes, the engine reaches its operating temperature, which it should ideally maintain under all operating conditions. The cooling, therefore, must be efficient but not excessive.

Reaching operating temperature rapidly is always advantageous. Unlike designs of a few years ago, it is now common to diminish the quantity of liquid in the cooling system, but to make it circulate more rapidly.

advantages in terms of noise suppression and can provide an accurate "thermal control" of the engine. The cooling system has a centrifugal pump to circulate the coolant, directing it through openings in the head and in the cylinder block and on to the radiator. The radiator is a water/air heat exchanger composed of two water tanks separated by a matrix of heat exchangers composed of a series of tubes and fins that increase the surface area for thermal exchange. If the tanks are arranged one atop the other, the radiator is vertical flow; if they are arranged to the sides of the matrix, it is cross-flow. The material used is an aluminum alloy that combines reduced density with high thermal conductivity. The matrix of exchange is crossed by air, which draws heat from the liquid in the tubes. Some motorcycles have two radiators, one on each side of the bike (a typical arrangement on off-road bikes).

■ Some motorcycles have two radiators, narrow and tall, located on either side of the frame's front downtube (Kawasaki).

■ Water-pump impellers have a series of radial blades that are straight or curved according to the model.

This is a diagram of the cooling system in a modern four-cylinder high-performance engine. The radiator is a cross-flow and has a curved shape (Suzuki).

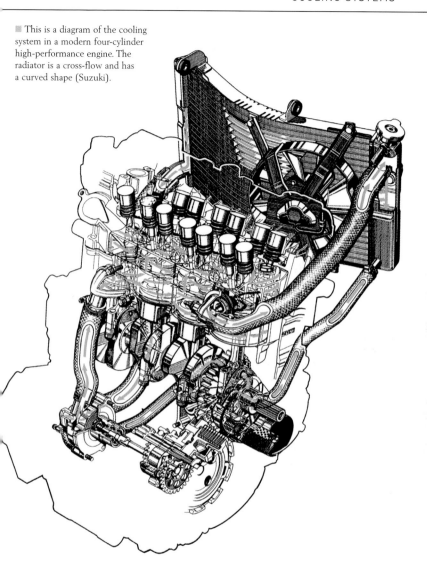

## Variations

Water pumps are usually positioned to the side of the crankcase. In some cases, however, they are fixed externally to the head and are driven by the end of the camshaft or by a shaft off the camshaft. Such an arrangement is appealing since it makes it possible to reduce the length of the couplings and the water lines.

To aid the engine in getting up to operating temperature quickly and to maintain constant temperature, the cooling system includes a thermostat that gradually opens to add coolant to the system as the temperature rises, beginning to do so at a specific temperature. Below that temperature, the cooling liquid that comes in contact with the walls of the head and the cylinder block cannot pass through the radiator, and thus heating takes place more quickly. Modern thermostats, equipped with two valves, are able to perform a "mixing" function and to contribute to excellent thermal control of the engine. Behind the radiator is an electric fan (sometimes there are two) that turns on only when the temperature of the liquid exceeds a certain point, a little below that of boiling. Usually the cooling system is completed by a recovery tank, connected by tubes to the radiator. Modern cooling systems are designed to function at high internal pressures.

Significantly, the water pump has a capacity on the order of 20 to 50 liters an hour for every unit of horsepower supplied by the engine. Most of the coolant (about 70 percent) is sent to the head, which in four-stroke engines absorbs far more heat than the cylinder. The opposite is the case with two-stroke engines, in which the cylinder is the most critical component for the cooling system. To avoid the danger of

In this engine, the water pump is mounted on the same shaft as the oil pump, activated by the sprocket of the primary transmission (Suzuki).

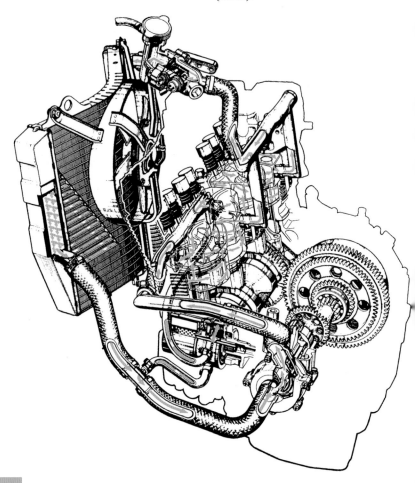

■ A four-cylinder engine with mixed air-oil cooling (Suzuki)

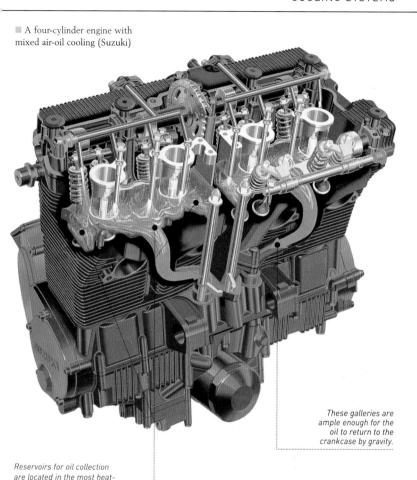

*These galleries are ample enough for the oil to return to the crankcase by gravity.*

*Reservoirs for oil collection are located in the most heat-stressed areas of the head.*

deformation, a particularly efficient cooling system is used in two-stroke competition engines, with the temperature of the water, at speed, far lower than the temperature reached in the cooling systems of four-stroke engines. In four-stroke engines the cooling system is focused on the head, with the goal of obtaining the best possible uniformity in heat distribution. In fact, the head has a complex shape and absorbs a great deal of heat and tends to have a "hot" side (the one with the exhaust valve and tube) and a "cool" side.

Additionally, modern engines use oil jets aimed at the bottom of the piston to improve piston cooling in high-performance four-stroke engines.

# LUBRICATION

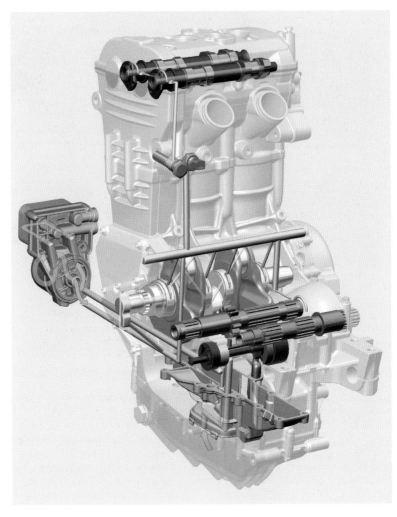

When two bodies are in contact, their reciprocal movement is resisted by friction, that is, a force that offers resistance to the movement. If the two bodies are not moving, starting to move one of them involves overcoming static friction and thus usually requires more force than if the body were already in motion. If the body is already moving, the force needed to keep it in motion, known as kinetic friction, is usually

■ The lubrication system in a BMW 800

much less. Whether initiating or maintaining movement, both of those cases involve sliding friction, and the amount of force involved will depend on the nature of the materials in contact and by the force pressing them against one another (aside from the conditions of the surfaces). The coefficient of friction

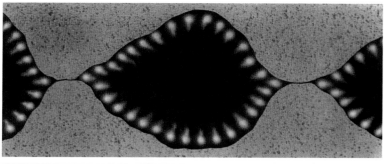

■ Thanks to the lubricating power of oil, polar molecules adhere to microscopic surface asperities, shown here enormously enlarged (top). This reduces the coefficient of friction and protects, at least in a good measure, the metallic surfaces in cases of boundary lubrication (Motul).

is the relationship between the amount of force resisting the movement (parallel to the surface of contact) and the amount of force pressing the bodies together (perpendicular to the surface). Its value varies according to the materials used. As the coefficient of friction increases, the resistance to movement increases. Reducing friction, which resists movement and "absorbs" mechanical energy transforming it into heat, is accomplished by lubrication.

In the case of sliding friction, various substances, but primarily oil, grease, and graphite, are introduced between the contact surfaces of the moving parts. No matter how smooth these surfaces seem they invariably have a certain degree of roughness, even if only the microscopic, raised solid features known as asperities. These form—on a vastly reduced scale, of course—a sort of mountainous landscape with peaks and valleys. Only the most accentuated of these microscopic peaks will touch, thus the surface area

in contact between the two bodies will be far less extensive than appears. The amount of surface roughness will depend on how the surfaces were manufactured and treated.

**Lubrication Regimes**
Deficiencies in lubrication inside an engine can cause mechanical breakdown; the overheating caused by friction leads to expansion, wear between parts, and in the most serious cases can cause mechanical seizure. Proper lubrication involves the introduction of a suitable fluid (always oil in engines and gears) between the surfaces of mechanical parts. The modes of lubrication, known as "regimes," differ according to the amount of fluid applied, which is related to the amount of resistance to movement. Various properties of the lubricating fluid are important, such as viscosity as is the speed at which the surfaces are moving. If there is extensive contact between the microscopic asperities on the respective surfaces,

■ Bushings work in hydrodynamic regimes and require a continuous and copious flow of oil.

■ Antifriction bearings, such as roller bearings and ball bearings, are less demanding, requiring only a "cloud" or splash of lubrication.

a thin film of oil will not be able to support the heavy load and the result will be what is called boundary friction. In this case resistance to movement is considerable, and the parts might be subject to wear. The thickness of the film of lubricant increases as the load diminishes and also as the viscosity or the speed increases. In a case where only some of the most accentuated surface asperities come in contact, the situation is one of mixed friction. Resistance to movement is far less and the amount of wear is negligible. When the thickness of the oil film increases in this situation, and it reaches a point

such that there is no contact between the surface asperities, the lubrication regime is called hydrodynamic and involves the "floating" of one component over the other. In this case, resistance to movement is minimal (it depends on the viscosity of the fluid), and wear is theoretically nonexistent. Adequate lubrication is absolutely indispensable, in particular for components such as those of the engine that work at high speeds and in the presence of considerable loads. For mechanical parts like main bearings and big-end bearings, which operate in a hydrodynamic lubrication regime, a copious and

## Oil Qualities

Engines are lubricated using specially formulated oils that are composed of a base and a "package" of additives that can account for more than 15 percent of oil's volume. The base oil can be mineral or synthetic. Mineral oils are refined from crude oil by refining procedures (atmospheric or vacuum distillation, hydroscission, and hydroisomerization) and are composed exclusively of hydrocarbons. Synthetic oils are composed of various kinds of chemical substances, the most common of which are polyalphaolefins (synthetic hydrocarbons) and esters, produced using a variety of chemical processes.

The additives are quite numerous. Among the most important are

detergents and dispersants, which prevent the formation and accumulation of varnish and sludge. There are also antiwear agents, enhancers of oiliness, corrosion inhibitors, and agents that improve viscosity. While the bases are somewhat similar, the additives differ greatly according to whether they are made to lubricate two- or four-stroke engines. Oils made for four-stroke engines must circulate in the engine and thus continue to perform over thousands and thousands of miles; whereas, oils for two-stroke engines do not circulate since they operate in a total-loss system and in fact enter the combustion chamber, where they burn up, leaving a negligible deposit.

continuous supply of oil is absolutely vital. In certain cases, instead of a fluid, rolling bodies are introduced between the two surfaces, thus transforming sliding friction into rolling friction, which is far inferior. Ball and roller bearings function in this way and require little in the way of lubrication (primarily oil to diminish heat).

## Grades of Motor Oils

The most important properties of motor oils are viscosity, lubricity, and the viscosity index (VI). Viscosity refers to the internal friction of the fluid itself, meaning its resistance to flowing. Its importance resides in the fact that it determines in a fundamental measure the load capacity of the film of oil. It must be adequate to the needs, but also not excessive so as to avoid a considerable absorption of power, resulting in a decline in the mechanical output.

The viscosity of the motor oils used in engines is rated on a scale developed by the Society of Automotive Engineers (SAE). Lubricating power assumes particular importance in the case of boundary lubrication. It indicates the capacity of the oil to adhere to the metallic surfaces, forming a thin film that is resistant to friction and extremely "slippery." Viscosity diminishes with increasing temperature following a progression that can vary considerably from oil to oil and that is indicated by the oil's VI. The higher the SAE number, the greater the oil's viscosity and the less variation there will be with changes in temperature. Modern motor oils are all multigrade (and thus have a high viscosity index). In other words, they have one SAE gradation at low temperatures and another at high temperatures (indicated on the container by two numbers). For example, an SAE 10W/40 behaves like an SAE 40 when cold and like a unigrade SAE 10 when warm. This means that at low temperatures it has

an elevated fluidity and that at high temperatures it nonetheless maintains good viscosity (indispensable to supporting high loads).

Together with viscosity, oil labels also give information on the quality, or "performance," level. This is indicated by two letters preceded by the letters API (American Petroleum Institute, the organization that established the tests for lubricants). For oils designed for use in four-stroke gasoline engines the grade is composed of an S followed by another letter, currently either M, L, or J. An SM oil is at a higher quality level than an SL, which in turn is superior to an SJ.

Other quality ratings have been developed by other organizations as well as by various automobile manufacturers. Some of these are based on especially severe quality and performance standards. Since the oil in motorcycle engines in general lubricates also the transmission and the clutch, it must be of an adequate type, meaning it must not contain such additives as friction modifiers, which are instead used in various lubricants made for use in automobile engines. For this reason there are two large categories of motor oil, indicated by the letters MA (without the additives and suitable for all four-stroke motorcycle engines) and MB.

## Lubrication Systems

In four-stroke engines suitable quantities of oil are circulated to various engine parts to lubricate them as needed. The lubrication system can be wet sump or dry sump. The "sump" is the lowest part of the engine's crankcase and the area where engine oil collects.

In wet-sump lubrication an oil pump circulates oil under pressure from the oil pan (or sump) to the different engine components by way of oil galleries. Gravity returns the oil to the sump after it has done its work where it collects

## Two Lubrication Systems

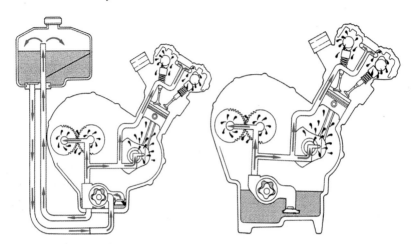

The dry-sump type (left) has two pumps: the pressure pump, and the scavenge pump. In the wet-sump system (right) the oil is collected in the oil pan, and there is only one pump (Honda).

and cools before being circulated again by the pump. The lubrication system includes a filter that retains even very small impurities. The filter is located below the pump and must be replaced periodically. There is also a mesh filter located near the oil pump intake tube that serves to catch larger impurities.

Dry-sump-lubrication systems have a second pump, called the scavenge pump, in addition to the pressure pump used in the wet sump system. The scavenge pump draws the oil from the sump in the lower part of the crankcase and sends it to a separate tank, sometimes located inside the frame tubes. From that tank the lubricant is drawn by the pump and sent into the lubrication circuit.

The dry-sump system makes it possible to reduce the vertical bulk of the engine. It can also diminish, at least to a certain degree, the power loss from oil sloshing and makes it possible for the pressure pump to draw the oil without interruption, even in the most challenging conditions (jumps, wheelies,

■ Oil reaches the connecting rod bushings under pressure, by way of openings in the crankshaft. The holes in the crankpin have rounded edges.

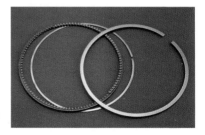

■ Oil-control rings remove lubricant from cylinder walls, leaving only a minimal quantity.

■ This diagram shows the wet-sump lubrication system in a modern inline four-cylinder engine (Suzuki).

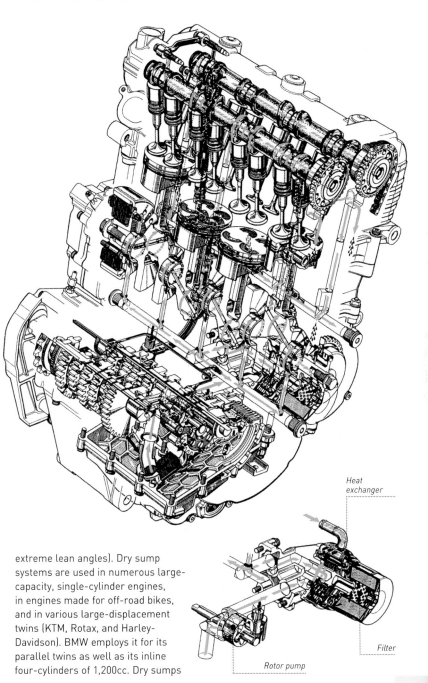

Heat
exchanger

Filter

Rotor pump

extreme lean angles). Dry sump systems are used in numerous large-capacity, single-cylinder engines, in engines made for off-road bikes, and in various large-displacement twins (KTM, Rotax, and Harley-Davidson). BMW employs it for its parallel twins as well as its inline four-cylinders of 1,200cc. Dry sumps

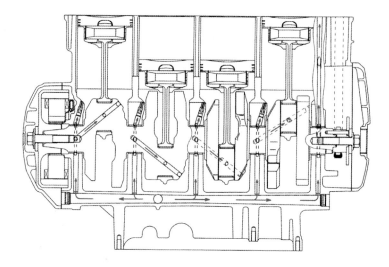

■ A diagram of the arrangement of the oil galleries in a crankshaft in which oil reaches the connecting rod bearings by way of the main journals (Yamaha)

are also used in MotoGP engines, most of which incorporate a separate oil reservoir in the crankcase in a cavity beneath the gear housing.

The wet-sump system offers the greater simplicity. It is interesting to note that several high-performance engines have only a pressure pump, but in effect the system of lubrication could be called "semi-dry" since the oil is collected in a sump separate from the crankcase.

### Pressure and Temperature

Two kinds of pumps are used in the lubrication systems of modern motorcycle engines: gears and rotors (trochoid). Gear pumps are composed of two closely meshed gears, a drive gear and a driven gear, mounted on parallel axes and set inside a fitted body. As the drive gear turns the driven gear, oil is "transported" by being forced between the gears to the other side. Rotor pumps are composed of two not-coaxial rotors, one set inside the other (the first has a smaller number of lobes); the oil is picked up as the inner rotor moves

around inside the outer. Oil is squeezed between them and thus moved from one side to the other.

Pumps with internal gears are widely used in today's automobiles but have only rare application in motorcycle engines. The pressure of the circuit is related to the output of the pump and the viscosity of the oil (and is thus greater when cold); pressure is primarily a result of the resistance the oil encounters inside the circuit. In engines with plain brass bushings, the "space" between them and the crankshaft journals is greatly reduced; the oil

■ Geared pumps are simple and economical.

## Rotor pumps

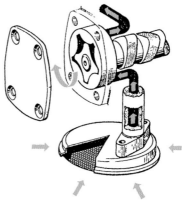

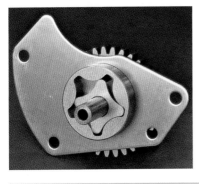

For many years, rotor pumps have been the most common application in motorcycle engines. Their design is clearly visible in these images.

encounters strong resistance to its passage and thus the pressure in the circuit is somewhat high. In engines with roller bearings on main journals and connecting rod bearings, there are large spaces between the flywheel bodies and the bearing journals through which the oil can pass; as a result, the pressure in the circuit is much less.

The lubrication system has a pressure-relief valve that opens when the pressure in the circuit rises above a certain level. Were the pressure in the circuit to reach excessive levels it could put parts of the engine in crisis, such as seals (and the pump itself), while also causing an absorption of power unacceptable to the pump itself. When the pressure relief valve opens, it bleeds off a volume of oil into the sump. Under certain conditions, this volume

can be quite high (even greater than the amount moving through the oil circuit's main galleries). To keep the oil in highly stressed engines from reaching

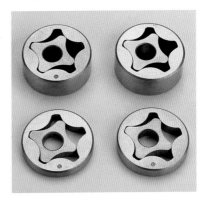

■ The rotors of a scavenge pump (above) are thicker than those of the main pump because of the greater oil volume they must handle.

75

■ The oil pressure relief valve serves to keep the oil in the circuit from reaching elevated pressures.

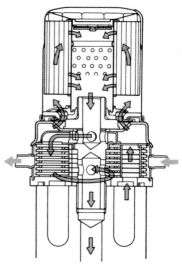

■ Water/oil heat exchangers are compact and often come incorporated in the cartridge filter.

excessive temperatures, radiators are used or water/oil heat exchangers. The heat exchangers are slightly less efficient, but have the advantage of being compact. Furthermore, to a certain degree, they help the engine reach operating temperature. (The water warms before the oil and thus contributes to bringing the engine to optimum temperature more rapidly.) In almost all high-performance engines, the oil also serves an important cooling function, for example the oil jets aimed at the piston bottoms.

### Two-stroke Engines

For many years, two-stroke engines have been lubricated by mixing a certain volume of oil with the fuel. In other words, these engines are not fed pure gas but rather a gas/oil mixture. Using this system, a true "cloud" of oil forms inside the crankcase and is deposited on the metal walls, forming a thin lubricating film. The system is adequate if roller bearings are used on the main bearings and in the connecting rod bearings, as these components are not excessively demanding in terms of lubrication. Later, at least in terms of motorcycles made for road use, oil-injection lubrication came into use.

This system draws the oil from a separate tank and sends it to the engine in quantities accurately measured on the basis of engine rotation speed and throttle setting (but typically in modest volume). In general the oil is injected directly into the intake port, immediately beyond the carburetor. Whether the oil is pre-mixed or injected, it's a "total loss" lubrication system, meaning there is no circulation of oil. The oil leaves the crankcase, passes through the transfer ports, and ends up in the upper part of the cylinder, where it is mostly burned together with the fuel mix. (A certain quantity leaves the exhaust without being burned.)

In most four-stroke motorcycle engines, the same oil lubricates the engine as well as the primary transmission and the gears. This is not possible in two-stroke engines, so separate oils are used. Given that they must work under very different conditions, the oils for two-stroke engines have different qualities than those used in four-stroke engines.

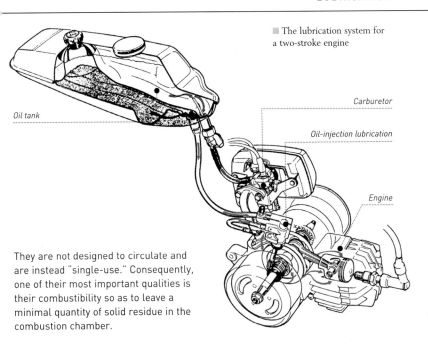

■ The lubrication system for a two-stroke engine

*Carburetor*

*Oil-injection lubrication*

*Engine*

*Oil tank*

They are not designed to circulate and are instead "single-use." Consequently, one of their most important qualities is their combustibility so as to leave a minimal quantity of solid residue in the combustion chamber.

■ Oil-injection systems deliver oil volume on the basis of engine speed and throttle opening.

# Mechanical Parts

# TECHNOLOGY
# AND MATERIALS

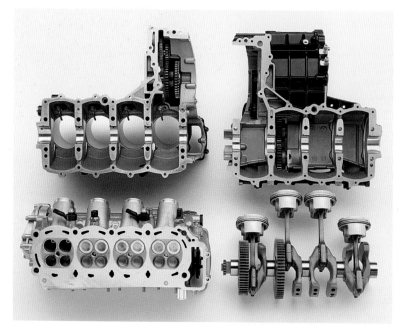

Motorcycle engines, transmissions, frames, suspension parts, and brake systems are made of metallic materials. Only certain "accessories," such as the sealing elements, flexible tubing, belts, and spark plugs are made of non-metallic materials. Before examining in detail the mechanical parts and the various devices that compose the motorcycle, it is useful to discuss the various metals used and their respective qualities. From there we can move on to a general description of production techniques and a discussion of basic mechanics.

Unlike electrical devices, motorcycles do not generally involve the use of pure metals, because such metals do not possess suitable properties. Instead, most motorcycle mechanical parts are made of alloys, which are created by combining a base metal with another

■ The mechanical parts that combine to make a motorcycle engine are made of materials with differing physical characteristics manufactured using a variety of processes (BMW).

element or elements. In addition to physical properties, every material has mechanical properties, the importance of which can be absolutely fundamental in terms of the variety of uses to which the material can be put.

**Physical Properties**
The density of a substance is the ratio of its mass to its volume (for which reason it could be called its volumetric mass). It is generally expressed in kg/dm³ or g/cm³. Given equal-size pieces of two materials, the material with the greater density will weigh more. Among the metals used for motorcycle components,

■ Today's most advanced trellis frames are made with chrome-molybdenum steel. Most swingarms are made from aluminum alloy (KTM).

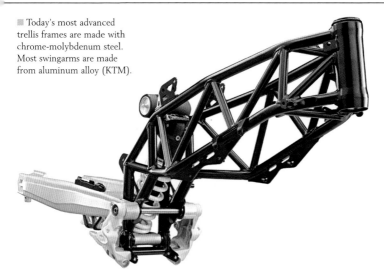

low density is found in both aluminum (2.7 kg/dm³) and, most of all, magnesium (1.8 kg/dm³), but these (in particular magnesium) do not have high mechanical qualities. By comparison, iron has a density of 7.8 kg/dm³ and titanium 4.5 kg/dm³.

Thermal conductivity is a measure of a substance's ability to conduct heat.

The higher the conductivity the more easily heat passes through the material. Thermal conductivity is expressed in W/(m °C). Both copper and aluminum have high thermal conductivity.

The coefficient of thermal expansion indicates how much a given quantity of a material will expand as the result of a 1-degree Celsius increase in

■ For some components, such as the crankshaft and piston connecting rods, rigidity and mechanical strength are the most important qualities, while for others, such as the pistons themselves, lightness and the ability to dissipate heat are more important (Yamaha).

■ One-piece crankshafts are forged from high-strength steel, usually alloyed with nickel, chrome, and/or molybdenum.

temperature. The higher the coefficient of thermal expansion is, the greater the expansion that the material will experience when heated. Thermal expansion is expressed in $°C^{-1}$. Aluminum has a high coefficient of thermal expansion, nearly twice that of iron.

### Mechanical Characteristics

When force is applied to a given material it may undergo deformation. For example, a material subjected to a tensile force may undergo a certain degree of lengthening. In engineering terms, this deformation is called elastic if it is reversible, meaning the substance returns to its original length when

the action of the force ceases. If the substance does not return to its original dimension, meaning it maintains even a small permanent deformation, the deformation is said to be plastic.

Resistance to traction indicates the breaking load, meaning the maximum force that the material can support before breaking.

When subjected to traction, materials behave in a variety of ways. Some undergo elastic deformation up to a certain point after which they may fracture; others (most materials) undergo an initial lengthening of the elastic type, after which they undergo a considerable amount of plastic

■ Built-up crankshafts are often made using cementation steel (obligatory for the crankpin, on which roller bearings must work, and for the connecting rod).

■ Pistons are made in aluminum alloy because of that metal's low density and high thermal conductivity (KTM).

■ Many aluminum alloys have excellent casting properties, making them highly suitable for the creation of geometrically complex parts made by casting, such as heads (Kawasaki).

■ Crankcases are subject to a good deal of stress but work at lower temperatures than those seen in the head. In this case, an aluminum alloy is the most logical choice.

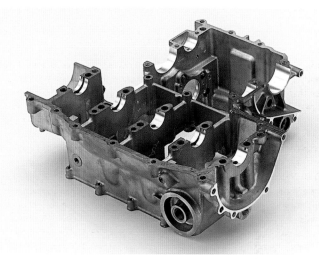

■ The two halves that typically constitute a motorcycle crankcase engine have neither undercuts nor internal cavities and can thus be made by diecasting (Kawasaki).

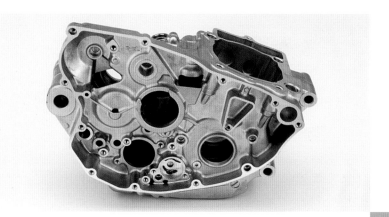

deformation before reaching the breaking point. The passage from the range of elastic deformation to plastic deformation is called the elastic limit, by which is indicated the quantity of traction the material can experience before undergoing plastic deformation. For some materials this coincides in practice with the yield point and, like resistance to traction, is expressed in N/mm². The ductility of a material is "measured" by the percent of permanent lengthening that a piece of the material undergoes when subjected to traction, before reaching the breaking point. The greater the lengthening, the greater the plasticity.

A substance with high resistance to traction and notable ductility is said to be high in tenacity. It is able to absorb a great deal of energy, deforming before breaking.

The elastic modulus indicates how much a substance deforms elastically under the action of a given force. The greater this modulus, the higher the rigidity. It is expressed in N/mm² or in GPa (gigapascal; 1 GPa = 1000 N/mm²). Hardness is a very important characteristic. It can be defined as the resistance a material gives to penetration or scratching. Very hard materials are indispensable for making engine parts with a high resistance to wear, whether that wear comes from sliding or from abrasion. In the presence of high levels of contact pressure, an elevated hardness in component surfaces is essential to allow correct operation and to ensure a long component life. (Such is the case, for example, with roller bearings.) Hardness can be measured by a variety of methods, most of which call for the use of an "indenter" with a spherical or pyramidal tip. Measuring the depth of penetration under standardized conditions, or calculating the size of the area of penetration, makes it possible to establish hardness. Hardness is measured by various scales, each of which gives a different unit of

■ Some frames with doubled beams are built up using sheet parts shaped and then welded together. Aluminum alloys are used, alloyed typically with magnesium or magnesium-silicon (Aprilia).

■ The connecting rods in race engines are often made from a titanium alloy, because it combines outstanding mechanical qualities with a far lower density than that of steel.

measurement. The Brinell method is usually used for materials that are not very hard, such as aluminum alloys, while for others the Vickers system or the Rockwell system is applied.

Impact resistance is a "dynamic" material quality indicating how well it resists blows. In other words, it expresses the level of fragility when a material is subjected to impact. It measures the amount of energy (with reference to a standardized section) needed to break a material sample. Impact resistance is measured in J/cm².

## Metallic Materials

The engine components that undergo the greatest mechanical stress are made from materials like steel and, in some cases, titanium. Where light weight is needed, use is made of materials that are less dense, such as aluminum alloys, even if their mechanical qualities are inferior.

Steels are alloys of iron and carbon in which the amount of carbon is less than 2 percent. Their characteristics are closely related to the percent of carbon, the quantity and type of other elements present, and the thermal treatment given the individual pieces. Steels thus form a large family of materials with highly differing qualities and costs. Those used inside engines can be divided into steels made by case hardening (for pieces that must have elevated surface hardness while also maintaining a high degree of internal

tenacity) and hardened and tempered steels. Within these two groups are the carbon steels (less expensive) and the alloy steels (with superior qualities). Particularly valued are steels alloyed with nickel, chrome, and/or molybdenum. Steels have a density on the order of 7.8 kg/dm³, a very high elastic modulus (more than 200 GPa), and can have resistance to traction superior to 1200 N/mm². Most steel components are made by forging. Engine parts that are usually made of steel include the crankshaft, connecting rod, piston pin, rocker arms, pushrods, roller bearings, and chains.

Aluminum alloys have a density on the order of 2.7 kg/dm³ and an elasticity modulus of about 70 GPa. They are valued for their high thermal conductivity and can be divided into two types based on the manufacturing procedures they undergo: cast alloys or wrought alloys (forging, extrusion). By far the most common of the cast aluminum alloys are silicon alloys.

■ Valves are made using special steels or super-alloys based on nickel (though only for exhaust valves in this latter case). The valves in some high-performance engines are made of titanium alloy, which results in a weight reduction (Kawasaki).

■ For some nonstructural parts, such as valve or crankcase covers, magnesium alloys are used for their greatly reduced density.

The presence of silicon imparts high fluidity, reduces the coefficient of thermal expansion, and contributes to improved mechanical qualities. Aluminum is used to make heads, cylinders, crankcases, and wheels, as well as frames. The alloys used for pistons form a category all their own, although they generally have high percentages of silicon along with additional elements.

The wrought aluminum alloys are divided into series based on the principal alloy element. Each member of the series is given a four-digit number. Some of these alloys are subjected to thermal hardening treatment. To improve the mechanical qualities of others, at least to a certain degree, work hardening (cold plastic deformation) is employed. Some wrought aluminum alloys do well when welded, while others do not, in which case the various parts are joined with rivets or glue, as is often done in aeronautical construction.

Aluminum-copper alloys have long been popular. Such alloys constitute the 2000 series. They were formerly referred to as Duralumin but are now better known in Europe by the commercial name Avional. Outstanding among the work-hardened aluminum alloys are those known as Peraluman, part of the 5000 series, in which the principal alloy element is magnesium. Also widely used are the aluminum-magnesium-silicon alloys of the 6000

series (Anticorodal). The best mechanical qualities are reached in aluminum-zinc alloys, which constitute the 7000 series and include Ergal, whose resistance to traction can reach 600 N/mm$^2$. The alloys of these two series can be hardened and are primarily used to make frames, forks, steering brackets, and wheels.

Magnesium alloys, which have an extremely low density (around 1.7 kg/dm$^3$), do not have elevated mechanical qualities and decline rapidly in heat. They are used for the manufacture of wheels for competition motorcycles, side covers, heads, and other parts not subject to elevated stress. The elastic modulus is only on the order of 45 GPa. Magnesium components are also easily corroded, making a suitable protective treatment necessary.

In recent years titanium alloys have become very popular. They have a notably lower density than steel (4.5 kg/dm$^3$ against 7.8) while their resistance to traction can be comparable. Their elastic modulus is on the order of 120 GPa. A titanium connecting rod can weigh 35 to 38 percent less than one made of steel. The only real drawback to these materials is their very high cost. Because of its expense, titanium parts (connecting rods and valves for example) are used exclusively on racing motorcycles and a few very high-end performance production bikes.

Cast iron (iron-carbon alloy in which there is more than 2 percent carbon) is used very rarely in modern motorcycles. It is used for some coated cylinder linings, for the seats and guides of valves, for piston rings, for the cylinders in scooters, and as the braking surface in drum brakes. Copper is present in the form of bronze (copper alloyed with tin) in certain bushings as well as in valve guides and seats.

## Manufacturing Processes

Various manufacturing technologies are used in making motorcycle engine components as well as other parts of motorcycles. Some parts are cast, a process that involves pouring liquid metal into molds, while others are forged, which shapes heated metal under pressure. Less common are parts

■ One-piece crankshafts for motorcycle engines are forged. Above is a rough crankshaft alongside one that has been finished and is ready to be installed.

made by sintering, which involves forming parts from powdered metal that is then heated. No matter the manufacturing process used, the rough pieces prepared must then be machine tooled to obtain final finishes that make them ready for assembly.

Casting requires metals that are readily castable in a liquid state. Obviously, the procedure is facilitated by using metals with low melting temperatures. Casting makes possible the production of parts with highly complex shapes. This is the procedure most often used in the production of aluminum-alloy parts. When the number of parts being produced is not great, sand casting can be employed. In this case, a model made of wood, resin, or metal is used to create an impression in compacted sand. The internal cavities and the undercuts are

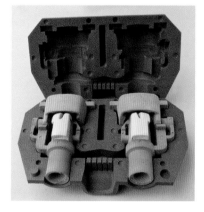

■ Above and below are shell molds used to make two-stroke engine cylinders. The photos show the two inner molds used for the casting.

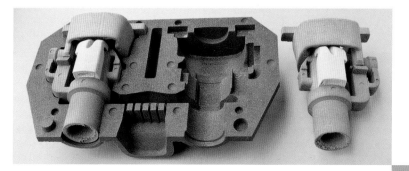

## Piston production process

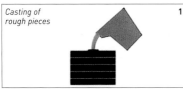

Casting of rough pieces **1**

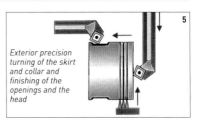

Exterior precision turning of the skirt and collar and finishing of the openings and the head **5**

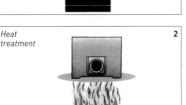

Heat treatment **2**

Finishing of the piston-pin hole **6**

Initial finishing and sizing processes **3**

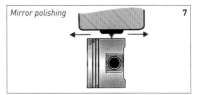

Mirror polishing **7**

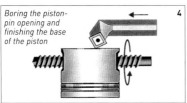

Boring the piston-pin opening and finishing the base of the piston **4**

Surface coating (for example, tinning) **8**

Final inspection, packing, and shipping **9**

These images illustrate the piston manufacturing process from casting through finished pistons.

obtained using cores in caked sand. The casting itself is done by gravity. There is also shell molding, a technology often used for motorcycle parts. This involves the creation of resin-covered "shells" into which liquid metal is poured. Another process, micro-fusion, makes it possible to create an impression by means of a wax model that later melts or is removed. A similar process involves making a pattern out of polystyrene foam and then packing it with sand. When molten metal is poured into the mold the foam vaporizes. Large-scale production often makes use of a different kind of shell mold in which the mold is made of metal and permanent, so that it can be used thousands of times. The flow of liquid metal into the mold usually relies on gravity, though sometimes the liquid metal is introduced to the mold using low pressure. The pieces obtained from these metal molds are superior to those made in sand molds since they are less porous, more compact, and possess a

■ The journals and lobes on the camshaft are worked using grind wheels to obtain excellent surface finishing.

finer crystalline "grain."

Diecasting is frequently used for large production series since significant quantities make it possible to amortize the high cost of the necessary tools. Diecasting involves injecting liquid metal into a multi-part metal mold under elevated pressure so that the mold fills very quickly, something on the order of tenths of a second. The system presents several limitations, chief among them poor mechanical characteristics as a result of notable porosity and the difficulty in creating internal cavities. These problems have been in large part overcome thanks to the recent development of very sophisticated and costly technologies, some of which involve the creation of a partial empty space inside the mold.

Forging involves the plastic deformation of a material at high temperature. In this way the metal, under the action of very high pressure, can "flow," assuming the shape of the mold. This is a costly production system, but it makes possible the creation of pieces with excellent qualities, and in certain situations it is the only method that can be used. (Certain materials do not have an adequate fluidity or melt only at very

high temperatures.) This technology is most often used for making steel components. Motorcycle crankshafts, for example, are all forged.

Sintering has less application in the field of motorcycle parts. Sintering involves the forming of objects through the high-temperature heating of compact metal powders. On occasion, parts made for prototypes, competition models, or particularly sophisticated models are made from whole metal, meaning machine tooled from metal plates, ingots, or bars. Such is the case, for example, with the bottoms for upside-down forks and with certain steering parts.

**Mechanical Tooling**
The rough pieces obtained through casting or forging must be subjected to a series of machine-tool processes to remove the excess metal and work the parts (journals, housings for bearings, sliding surfaces, or seal elements) to their final size. These processes also make it possible to achieve the desired final shape and to give the surfaces their proper finishing. Several stages are often involved in these processes. For example, in the case of crankshaft journals, the work begins with turning to achieve the "rough shaping," which is

■ Transmission gears need extremely hard teeth and work surfaces while at the same time having great internal tenacity, meaning that nonsurface areas cannot undergo the hardening process. Precision treatment is essential (BMW).

■ The cylinder liners of most modern motorcycle engines have coated exteriors (a matrix of nickel with particles of silicon carbide) applied galvanically.

then followed by finer turning and then by finishing (a process done on a machine fitted with a grinding wheel). Some components must undergo a large number of different treatments. Such is the case with heads for which not only must the upper and lower sealing surfaces be worked with precision, but the housings must be made for guides and valve seats, as well as the surfaces through which the intake and exhaust ports will pass. Additionally there must be supports for the camshafts and the housings for the lifters, or tappets.

■ The working surfaces of cylinder rings (the area in contact with the cylinder walls) are given a thin coating of a material such as chrome or molybdenum.

### Treatments and Coatings

In some ways, the manufacturing procedures can be traumatic for the components. In particular, a good deal of internal tension is sometimes created, but this can be reduced or even eliminated by means of heat treatments. Final mechanical qualities are imparted to the various mechanical parts by way of these treatments, which serve a critical step in the manufacturing process. Some of them involve the entire mass of the metal while others involve only the exterior layers.

Heat treatment involves heating the piece to a certain temperature, keeping it at that temperature for a set period of time, then cooling it. The cooling process must be carefully controlled; indeed, the speed at which the piece returns to ambient temperature is of critical importance. The processes of hardening and tempering are of particular importance to steel components. Hardening involves bringing the piece to a very high temperature and then rapidly quenching it by immersion in water or oil. By way of this process, steels containing more than 0.2 percent carbon are given a greater hardness and higher resistance to traction. Tempering is used to reduce the brittleness induced by heat treatments and to improve strength and resilience. It involves heating the piece to a temperature much lower than that used for quenching, but then cooling it slowly. The two operations of hardening

■ In connecting rods made to work on rollers, the internal surfaces of the eyes require an elevated degree of hardness. Thus, they are case-hardened, taking care to protect the exterior surfaces from the treatment (these are often given a copper coating) to avoid making them brittle.

■ The lobes on camshafts often operate under critical conditions (high contact pressure, mixed if not limited lubrication). For this reason they are made very hard and are often given a protective anti-wear treatment such as a surface coating of phosphate (Yamaha).

and tempering are thus related.

Aluminum alloys are sometimes subjected to similar treatments but at notably different temperatures. Thermochemical treatments call for keeping the piece in an atmosphere particularly rich in elements such as nitrogen and carbon, at elevated temperatures, for a long period of time. In this way the steel is case-hardened: the surface layers "absorb" the elements in question. The most commonly used case-hardening methods are induction hardening, carburizing, and nitriding. In all cases the surface is rendered very hard while the inner part of the piece maintains great tenacity.

In many cases, the work surfaces of a component are given a particular hardness, smoothness, or resistance to wear. They can also be given a layer of protection. For these purposes coatings are used, which can be applied by galvanization or by spraying (flame-

spraying or thermal-spraying). The result is a thin layer of a material different from the material of which the piece is made. Among the best known coatings is chrome, used on rocker-arm pads, on piston rings, and on valve stems. Cylinder liners are often given a coating of nickel dispersed with very hard silicon carbide particles. The fork stanchions of some recent bikes have been given a coating of titanium nitride.

**Dimensional and Geometric Tolerances**
Though parts may be worked with care, measuring them with extremely precise instruments in the process, there will always be small deviations from the part's ideal dimensions. In the same way that absolute perfection is not obtainable, no two manufactured components will be exactly identical. Such deviations from the ideal constitute dimensional and geometric tolerances.

Engineers take this into account and

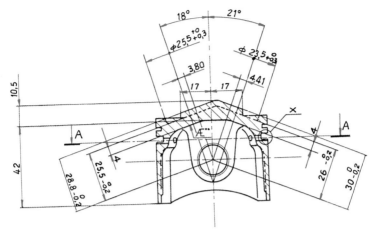

■ The critical dimensions on engine component blueprints are always accompanied by tolerances that must be respected, or by permissible limits of variation with respect to the "ideal" measurement (AE).

specify the acceptable "range" of deviation for every piece of each mechanical part, meaning they indicate in the design the maximum deviations admissible with respect to the measurements and the ideal forms.

Components can be fitted with clearance or with interference. Fitting with clearance means that after the components are mounted there is room between them for a certain degree of movement. A bearing journal, for example, has a diameter less than that of the opening into which it is inserted; as a result it is free to turn. Pistons are mounted with clearance in the cylinder liners, as are the valve stems in their guides and the lifters in their housings.

■ Two dial gauges are used to determine rotational errors of the two journals of a crankshaft. Ideally the deviation from spec should be zero.

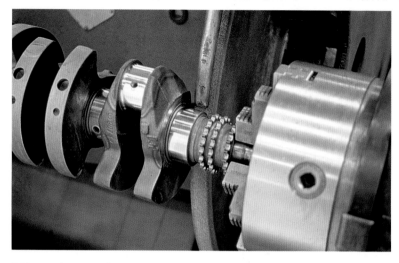

■ The main bearings and connecting rods for crankshafts are worked on grinding machines able to ensure minimal deviations from the theoretically perfect shape.

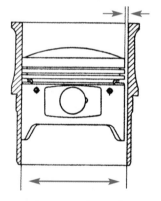

When two components are fitted with interference there is no room for movement; they are firmly joined. In this case, the journal would have a greater diameter than the housing in which it is inserted. The mounting must be done by use of force (an interference fit is also called a press fit) or by means of a thermal method. (The part with the opening is heated to make it expand, or the piece to be inserted is cooled.) The interference-fit components in a motorcycle engine include the crankpins in the flywheels of built-up

■ There must be a certain degree of diametrical clearance between the piston and the cylinder liner to permit the free movement of the piston and the formation of a film of lubricant on the cylinder walls (Lancia).

■ The rings in the valve seats are inserted in their housings in the head, with a degree of interference to ensure a firm fit and optimum contact between the surfaces.

The assembly of built-up crankshafts is done by inserting the crankpins through holes in the flywheels. The assembly calls for a high degree of interference, requiring the use of a press.

A modern motorcycle engine is made up of many components machined to very close size and shape tolerances and assembled with great precision (KTM).

crankshafts, the guides and seats of the valves in their housings, and the roller bearings in the walls of the crankcase. Such fits are firm and stable since the walls of the housing deform elastically and therefore tighten forcefully around the inserted element. Both clearance fits and interference fits must be calculated with extreme accuracy, taking into account the width of walls, the physical characteristics of the materials, the sizes of the parts, and the temperatures they will reach while functioning.

# ENGINE
# COMPONENTS

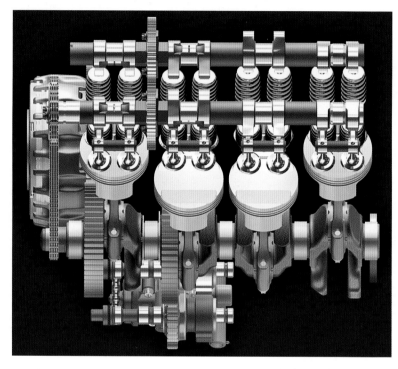

■ The arrangement of moving parts in a modern, high-performance inline four-cylinder engine (BMW)

### Engine Components

Engines are composed of many mechanical parts. Each is called upon to perform a specific function, and each part does so under varying degrees of stress, both thermal and mechanical. The fixed components are the head, cylinder, and crankcase, plus the various covers (crankcase side covers and valve covers). These parts are usually made of aluminum alloy. The mobile components are the pistons, connecting rods, crankshaft, and the valvetrain parts (valves, lifters, rocker arms, and camshafts). Some of these are in direct contact with gases that reach high temperatures during combustion and are thus exposed to high levels of thermal stress. Other parts experience strong accelerations and thus are highly stressed from a mechanical point of view. The design and manufacture of each of these components take these factors into consideration so that the components will have the highest levels of reliability under all operating conditions and will also have long useful lives. As for a motorcycle's overall life span, it is clear that the demands made on a touring bike or a normal street bike are far different from those made on a motocross or competition enduro bike.

## The Head

The cylinder (which for convenience we will imagine as being arranged vertically) is closed at the top by a removable piece called the head. In two-stroke engines this is a simple cover with the combustion chamber in its underside, channels for coolant (if water cooled), and cooling fins on its exterior. In four-stroke engines, the head has a far more complex shape, given that, aside from the combustion chamber and the channels for the coolant (or fins if air cooled), it must also house the intake and exhaust ports along with the valves and all the related parts that control their movement (camshaft(s), lifters, or rocker arms). Since most of today's motorcycle engines are overhead camshaft or dual overhead camshaft, the upper part of the head also carries the supports in which to insert the camshaft bearings. In most cases these supports can be

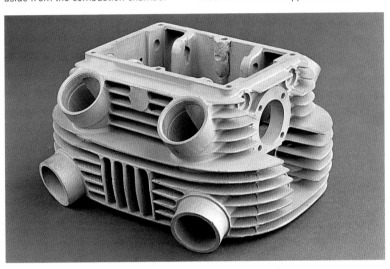

■ The heads for air-cooled engines have many fins. Visible in this photograph is the space for the camshaft and the rocker arms.

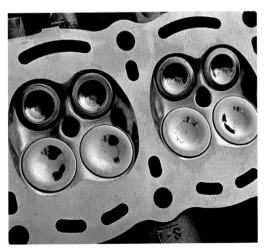

■ Combustion chambers are located on the underside of the head. Also visible in this photograph are the water passages and the central positioning of the spark plugs.

This is the combustion chamber in a competition engine. The ports are large and nearly straight; the valve seats are made of copper beryllium. The spark plug is at the center. In this case, it's a superficial-discharge plug.

This head of a high-performance single-cylinder engine is made for off-road use. This is for a single-camshaft valvetrain.

disassembled, and often their upper halves are cast directly in the valve cover. Since this cover is not subject to enormous stress, in some high-performance engines it is made of magnesium alloy. In certain cases the camshafts and sometimes even the lifters or rocker arms are housed in a removable housing; at other times there is a cambox bolted over the head, the structure of which is simpler.

The valve guides and seats are also resident in the head. Their housings must be machined with extreme care to ensure excellent contact and to provide the correct interference fit.

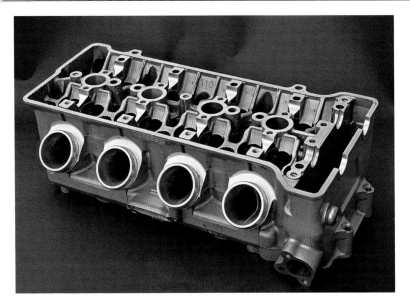

The head is a fixed component, but it is subject to a high degree of stress. Inside the cylinder the piston goes through its strokes toward combustion, which occurs once every two turns of the crankshaft in four-stroke engines. Each combustion stroke causes a spike in pressure, and the release of an enormous quantity of heat directed at the underside of the head and absorbed in the cylinder and head walls. This heat is drawn away by the cooling fluid in its channels. The walls of the exhaust port

■ This image shows the complexity of a head for a four-cylinder engine with high power output. The supports for the two camshafts are visible, along with the housings for the bucket tappets, the "wells" for the spark plugs, and the opening for the timing chain.

■ In order to reduce the structural complexity of the head it is sometimes divided into two parts; the upper part ("overhead") has the supports for the camshafts and the housings for the valve lifters.

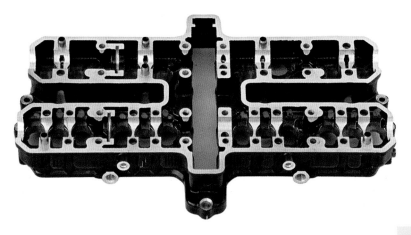

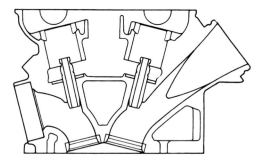

■ This is a side view of a head with twin camshafts. The intake and exhaust ports are visible along with the valve guides and seats (Suzuki).

also receive a great quantity of heat. Since the head also houses the intake ports it tends to have a "hot" side and a "cool" side. Obtaining the best uniformity in the distribution of temperature constitutes one of the most challenging and important aspects in head design.

These components are invariably made of aluminum alloy, which combines a modest density with elevated thermal conductivity. In general, the alloys used contain from 7 to 10 percent silicon, which aside from improving the casting properties gives the material

■ The heads of two-stroke engines have a very simple structure. The absence of valves and ports makes possible greater freedom in terms of the shape of the combustion chamber.

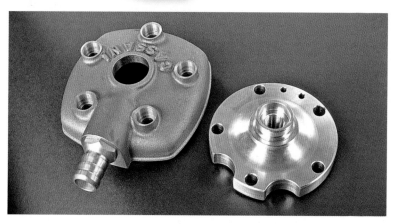

■ The heads of two-stroke engines are sometimes made in two pieces: a bowl that contains the combustion chamber and a cover mounted on top of it.

improved mechanical qualities and diminishes the coefficient of thermal expansion. The manufacturing process used for the head production is usually shell-mold casting using sand cores. Less frequently heads are made by way of casting in sand-resin molds or using the lost-foam procedure. Heads for two-stroke, liquid-cooled engines are sometimes made in two pieces. The inner piece is a simple bowl in which the combustion chamber is opened, and the exterior is an even simpler cover.

## Valves

The intake and exhaust ports bring each cylinder in contact with the external environment. (Fresh air is drawn in on one side, and the burned gases are expelled on the other.) Each port terminates internally, meaning in the combustion chamber, with a seat to which the valve returns when it closes. When the valve is closed and in contact with the seat, it prevents the passage of gases, and when open, meaning lifted off its seat, it permits the passage of gases.

■ The shape of a modern head for a very high-performance engine. Clearly visible are the valvetrain and ports. In this case, the cams act on rocker arms (BMW).

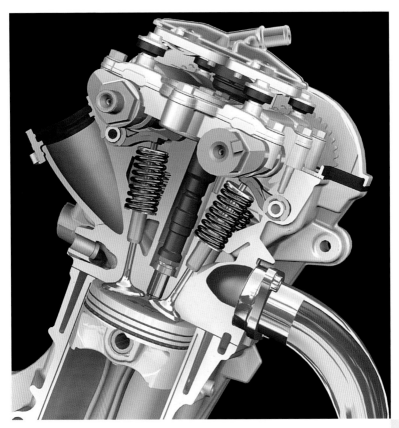

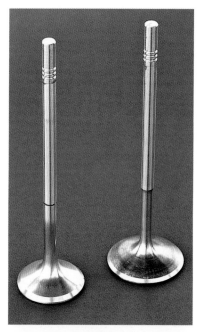

■ Although simple in shape, engine valves are complicated to make. Manufactured to withstand high levels of stress, they are authentic metallurgical masterpieces. In particular, the exhaust valves, which must withstand extreme conditions from a thermal point of view, require the use of sophisticated materials.

Valves are composed of a head (the part that acts as the stopper) and a stem. Near its end the stem has a small groove to which valve keepers are attached that, in turn, connect to the return spring cup. The outer surface of the valve head has a truncated-cone shape (with an angle of 45 degrees); this is made to rest against the similarly shaped surface of the seat to ensure a perfect seal when the valve is in the closed position.

Valves are subject to a high level of stress since they must support the pressure of the gases, and they are also subject to very elevated acceleration. When the engine is turning at 6,000 rpm, each valve must open and close 50 times per second; at 12,000 rpm, they open and close 100 times per second. Exhaust valves are exposed to the high-temperature gases that exit the cylinder while they are raised, a position in which they can dissipate only a limited amount of the heat they are absorbing. In fact, they are cooled by coming into contact with the valve seat, which conducts the heat away from the valve to the cylinder head. Valves function under very difficult conditions, to say the least; it is not rare for valve heads to reach temperatures over 800 degrees C.

■ Intake valves always have a larger diameter than exhaust valves. (This is true even if the engine has four valves per cylinder.) The larger size reflects the fact that it is harder to pull in the air/fuel mixture than it is to expel the burnt gases from the cylinder.

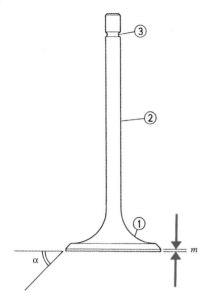

The situation is different for intake valves, since they are subject to excellent cooling by way of the fresh air/fuel mixture that comes through them. They operate at a comparatively low temperature, which means they can be manufactured with less "sophisticated" and less costly materials. Aside from the fact that two or four valves are used per cylinder (up to five valves in some designs), the intake valves are always larger in diameter than the exhaust valves. The reason for this is simple: Expelling burnt gases is much easier than drawing in the fresh air/fuel mixture.

Indeed, most of the burnt gases escape without being "pumped out" by the piston. Because of this the passageways used for intake are wider than those for exhaust. (The valves and ports are greater in diameter.)

Valves are made by hot, plastic deformation (forging and swaging) followed by precision machining. Special steels are used that contain

■ The parts of the valve are the head (1), the stem (2), and the groove (3) to which the keepers are connected. The *m* indicates the valve margin; the *a* indicates the angle of the sealing surface.

■ Comparison of the heads of two successive versions of a high-performance engine. Clearly visible is the angle between the valves, very reduced in both cases, and the spaces for the rocker arms (Kawasaki).

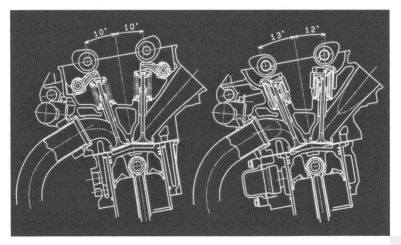

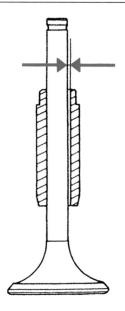

■ Between the valve stem and the guide hole there is a diametrical clearance on the order of a few hundredths of a millimeter (Lancia).

considerable quantities of chrome, manganese, or nickel. The exhaust valves in highly stressed engines are often made of nickel-based superalloys with high percentages of chrome; Nimonic and Inconel are two examples. In certain cases the valves are subjected to treatments of soft nitriding (with superficial enrichment of nitrogen obtained through immersion in a melted salts bath). Very often the stem is given

a thin coating of chrome. The area of the valve head that seals against the seat is often given a layer of stellite (an alloy of cobalt, chrome, and tungsten with excellent wear resistance as well as resistance to heat corrosion). Racing engines and those of some sportbikes are fitted with lightweight titanium-alloy valves.

*Valve Guides, Seats, and Springs*
Valve stems fit inside valve guides, small cylindrical elements often with a recessed lip, which is fitted to the head with a certain degree of interference. Valve guides are made in cast iron or bronze and typically have a small oil seal at the upper end to prevent the entry of lubricant between the guide and the stem. (There is a few hundredths of a millimeter clearance between the stem and guide.) The space of the guide must be perfectly concentric with the seat of the valve.

Valve seats are ring-shaped inserts fitted in the head with a high level of interference. The valve head rests on the seat's truncated-cone surface when the valve is in the closed position. The thermal method is usually used to fit valve seats into the head. The head is brought to a high temperature to make it expand and permit the easy insertion of the ring, which is at ambient temperature. The procedure can also be accomplished by cooling the rings (using liquid nitrogen or propane) to

■ In closed position, the valves do not rest directly on the aluminum of the cylinder head but rather on rings made of a hard material interference mounted in the head itself.

Helical springs are used to close valves. Those shown here, together with their cups, are dual-coil springs.

Very often, as here, valve springs have a progressive-rate coil.

make them contract. Valve seats are made of nickel-chromed cast iron, strongly alloyed steels, or even bronzes, often containing considerable quantities of aluminum, plus iron and nickel. The valve seats in racing engines are often made of beryllium copper with low percentages of cobalt or nickel.

The valve spring serves not only to return the spring to its seat (and hold it

there) but also to make certain for the entire period that the valve is open that the lifter or rocker arm remains in contact with the lobe on the camshaft. This is critical to ensure that the valve's movement takes place in precisely the way it was designed. Valve springs are made of steel wire with molybdenum chrome, chrome vanadium, or silicon with chrome. To increase the spring's fatigue-resistance, the spring

■ Camshafts usually have one cam lobe for each valve. Visible in this photo are the journals with their ground finish and the cams, which have been given a thermochemical treatment.

wire is shot peened after it is coiled.

Most valve springs are made with variable rate coils. Recently, various makers have adopted truncated-cone springs in place of the conventional cylindrical springs. The cone shape reduces the spring's mass and permits the use of small and lighter cups. Each valve is connected to its spring by the cup and the keepers. These are small steel parts that on one side are inserted in a groove of the valve stem and on the other are inserted in an opening in the cup (a kind of plate that presses against the upper end of the spring).

**Camshaft, Pushrods, and Rocker Arms**
The points at which the valves open and close, the height of their maximum lift, as well as the modalities with which their movement takes place are all determined by the shape and position of the camshaft's lobes. The camshaft's rotations must be correctly timed to correspond to the crankshaft rotation.

The rotations of the two shafts must be perfectly synchronized (at a ratio of 2:1), and the camshaft must be positioned in such a way that the valves open and close exactly at the correct time in order to obtain the valve timing established in the design. During their rotation, the cam's lobes make the part they come in contact with rise. In the language of engineers, these parts being acted upon by the lobes make a reciprocating movement. If the part is moved in a straight line, running up and down inside a cylindrical housing, it is a pushrod. If, instead, its movement describes the arc of a circle, oscillating on a fulcrum, it is a rocker arm. The parts of the lobe include a base circle, two flanks (one of opening and one of closing), and a maximum lift (or nose). To make the valve lift and return to its seat in a smooth and gradual motion, the flanks of the lobe are connected to the base circle by two ramps.

Camshafts are made in steel or cast

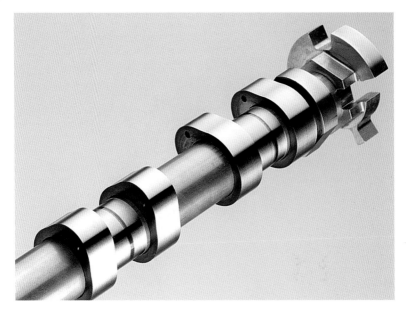

■ Some recent engines have tubular camshafts with thin walls and coated cams (BMW).

iron and are often surface treated to give increased hardness to the lobes' work surface. In the past this was almost always done by case-hardening, but today's preference is nitriding.

In engines with two overhead cams the lifters are always the bucket-tappet type and are made of steel. Engines with one overhead camshaft have rocker arms, pivoting centrally. (In some cases

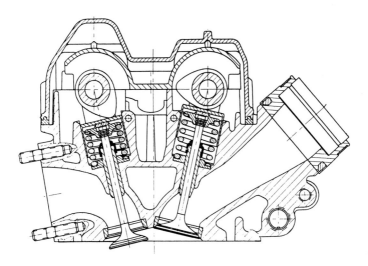

■ Here is a cross section of the double overhead cam of a modern V twin. Note the single springs, the wide-diameter bucket tappets, and the reduced angle between the valves.

■ In general, the drive sprocket has a specific position on the camshaft. Changing the sprocket position (for example, by way of slotted holes for the mounting bolts) changes the timing. This is a common technique used on competition engines.

the arm on the side opposite the cam is doubled since it operates two valves.) The rocker arms are usually made of forged steel and contact the lobe by way of a curved plate that is given a hard surface coating like chrome.

The movement of the valves is determined by the shape of the cam

lobes. Even so, this movement is also influenced by the shape of the rocker arms. (The parameters at play here are the radius of curvature of the small plate and the ratio between the length of the two arms.) Every lobe's shape must be determined by taking into consideration the shape and arrangement of the part on which it acts. The lift of the cam is not always identified with that of the valve. In some cases the rocker arm has a roller instead of a small plate by which it contacts the lobe. Finger-type rocker arms are pivoted at one end, making them true "rocking" levers. They are used in some high-performance twin-cam engines.

■ Shown above are valves with their lifters, known as bucket tappets. Note the tappets' simple shapes and thin walls.

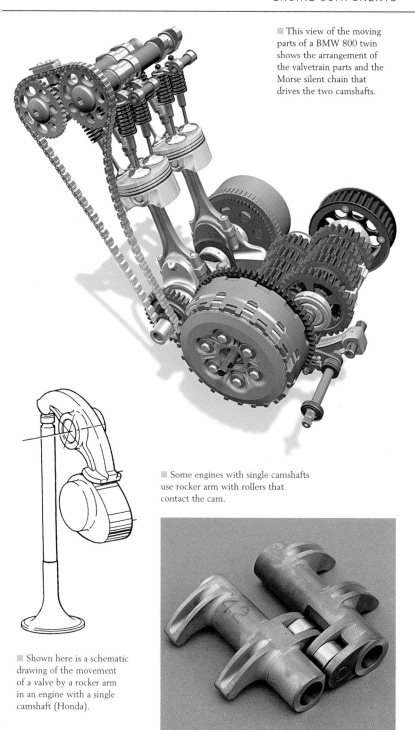

■ This view of the moving parts of a BMW 800 twin shows the arrangement of the valvetrain parts and the Morse silent chain that drives the two camshafts.

■ Some engines with single camshafts use rocker arm with rollers that contact the cam.

■ Shown here is a schematic drawing of the movement of a valve by a rocker arm in an engine with a single camshaft (Honda).

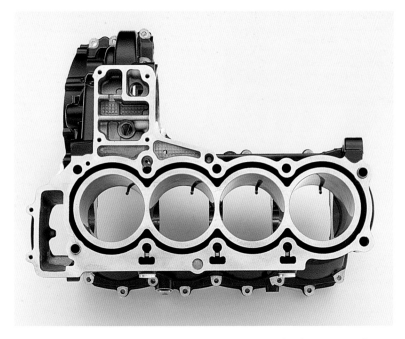

■ This is the view from above of a cylinder block that is incorporated in the upper part of a crankcase with an open-deck structure and Siamese cylinder liners (BMW).

### The Cylinder

The piston is housed in the cylinder, a component located between the crankcase and the head. Inside the cylinder is a liner, or sleeve, which can be integral (cast together with the cylinder) or it can be a coating applied using one of a variety of technologies. The piston is inserted in the liner with a diametric clearance on the order of a few hundredths of a millimeter; the seal between the two components is accomplished using rings. In order for the engine to function well while providing the best possible perform-ance, the cylinder lining must have a surface that allows the formation of a uniform, continuous film of oil that will make the rings slide under optimal conditions. The liner must also have the correct geometry. In other words, so far as is possible, it must be cylindrical and must maintain that shape in any situation. For this reason the

manufacturing tolerances must be especially strict, and the design of the cylinder must ensure that no distortions of an appreciable size result from the passage from the ambient temperature to the temperature of operation (it must not gall, meaning suffer damage as a result of friction), or following the tightening of the bolts that hold the head in place.

The material that surrounds the liner (the walls of which must have an adequate thickness) and even the position of the holes for the head studs can have considerable importance. It is of fundamental importance that the temperature be uniformly distributed. As for the geometric tolerances, the finished liner must not have an oval or upper cone shape greater than 100th of a millimeter. (Some manufacturers use an even smaller measurement for errors.)

Other possible errors (barrel shaped

■ Cylinders that are in direct contact with coolant are sometimes given coatings. The example here has an upper flange and is shown with its piston, the latter featuring a truly "extreme" design.

or hourglass shaped) along with cylinder wall lobations or waviness (which could be caused by vibrations during the machining process) must be reduced to a minimum. It is also indispensable that the liner's central axis be perfectly perpendicular to the crankshaft's axis of rotation. In engines with several inline cylinders, the axes of the liners must be perfectly parallel. The same applies to the liners of every cylinder bank in engines with more than two cylinders in a V configuration.

Additionally, the upper and lower surfaces of the cylinder must be completely smooth and parallel.

*Integrated or Removable*
In the case of traditional motorcycles, the cylinder (or the cylinder "block," if there is more than one and they are aligned) should be removable. In that case, it is cast separately and then bolted to the top of the crankcase. Long bolts (head studs) are often used, and these extend directly from the crankcase to the head, with the cylinder thus held between these two components. In other cases, however, the bolts that hold the head are independent of those that hold the cylinder to the crankcase.

■ Cylinders for air-cooled boxer engines. The fins are visible, as are the openings for the timing chains. The liners are integral, with coating on the work surfaces (BMW).

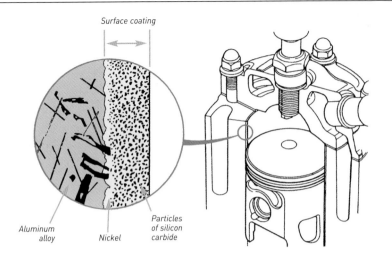

*Surface coating*

*Aluminum alloy*

*Nickel*

*Particles of silicon carbide*

■ Today, most cylinder parts are given a surface coating of nickel with silicon carbide. The image is based on a two-stroke engine made by Honda.

■ Cylinder block with closed-deck structure directly incorporated in the casting of the upper part of the crankcase

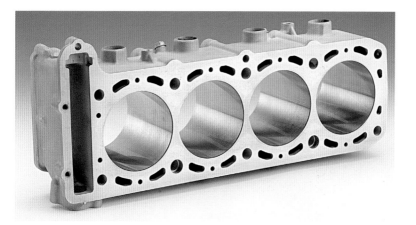

■ This four-cylinder block has a closed-deck structure but is removable, made from an integral casting (Kawasaki).

Recently, engines in which the cylinder block is not removable and is instead part of the upper part of the crankcase have acquired popularity.

The cylinders are made in aluminum alloy and are typically made by casting. Cast-iron cylinders are not used in modern engines but find discrete application in scooters and various small engines.

In air-cooled engines the cylinder's external surface is covered with rows of fins. In water-cooled engines there must be space for the coolant between the cylinder and the exterior wall of the cylinder block.

Since aluminum alloy does not possess adequate wear-resistance qualities, special technologies have been developed for the manufacture of cylinder liners. The most common applies a thin coating of galvanized nickel with particles of silicon carbide directly on the aluminum of the liner (which in this case is generally integral with the cylinder). The total thickness of the coating is about 0.05 to 0.08 millimeters. The particles of silicon carbide, which have a hardness rating of 2,200 to 2,500 on the Vickers scale, have a diameter on the order of 2.0 to 2.5 microns (thousandths of

a millimeter). In addition to providing excellent heat transmission and ensuring great wear resistance (and thus a long life for the component), a coating like this permits rapid seating of the cylinder rings.

**Wet Sleeves and Dry Sleeves**

In the past, the most common method of fitting liners involved inserting them dry, putting the liner in direct contact with the light alloy of the cylinder but not touched by the coolant. In this case, the liner was made of cast iron and was incorporated during the casting of the cylinder or installed in the cylinder with a certain degree of interference. Even today, some motorcycle engines still make use of this construction method. If incorporated by casting, the external surface of the liner usually has a series of projections that increase the area of heat exchange and ensure an excellent mechanical "anchorage." When installed with an interference fit, a technique used only rarely today, both the liner and its housing are given an raised finish (indispensable for the creation of intimate contact between the surfaces). The insertion of the liner (made of cast iron) can be effected with the use of a press or by the thermal method.

■ In open-deck cylinders, the side walls are separated from the upper part of the liners. This makes the casting easier (Yamaha).

Liners inserted with an interference fit can be replaced and often have a flange at the top.

Some engines use wet sleeves instead, meaning the liners are in direct contact with the coolant. In this case, the liners can be given an upper or lower flange. The first configuration, using liners with an upper flange, involves the use of a cylinder block with a closed-deck construction and is by far the preferred configuration. The second arrangement allows the creation of a cylinder block with an open deck with an extremely simple structure. Dependent upon the model, wet sleeves can be made of cast iron or aluminum alloy with a surface coating applied to the work surfaces.

Far less common are cylinder blocks in which the liners have been made by incorporating porous elements called preforms, composed of ceramic fiber with hard particles, in the casting. These preforms are thus literally impregnated in the liquid aluminum.

In cylinders with a closed-deck structure the sleeves are connected both above and below to the outer walls. In open-deck configurations there is no true upper surface on which the head can rest, and there is no direct connection between the uppermost end of the sleeve and the outer walls of the cylinder. This last method is favorable from a construction point of view but is slightly inferior when comparing examples of the same size, in terms of its structural rigidity.

**The Piston**

The piston is housed in the cylinder liner, and during operation of the engine it moves up and down, alternating from TDC to BDC and back again. In high-performance engines the piston is subject to high levels of stress. To begin with, it is subject to continuous acceleration and deceleration, given that at the two dead points it must stop and change the direction of its movement. In an engine running at 8,000 rpm the

■ The cylinder architecture in two-stroke engines differs from that of the four-stroke engines because of the presence of openings and ports. The photograph shows the openings for the transfer ports.

piston makes 266 strokes per second; in an engine running at 12,000 rpm it makes 400. The accelerations (and thus the forces of inertia) are enormous. The pressure of the exploding air/fuel mixture strikes the crown of the piston and does so with great rapidity and power. In the case of a piston with a bore of 80 millimeters, the force driven into it can amount to more than 4 tons. As for thermal stress, it should suffice to point out that the piston is directly hit by the exploding gas and a blast of enormous heat.

In addition to its principal function of moving up and down, which draws

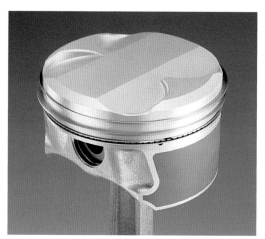

■ The upper surface of the piston (its crown) constitutes the mobile wall of the combustion chamber. The skirt, meaning the part below the rings, performs the function of guide (BMW).

In recent years, pistons made for high-performance engines have seen the progressive reduction of their height in relation to their diameter and the extension of the circumference of the skirt (Kawasaki).

in and expels the gases from the cylinder during the "cold" stroke and compresses them to then receive their thermal energy during the "hot" stroke, the piston has the important role of guiding the foot of the connecting rod, meaning it directs onto the cylinder walls the transversal force caused by the fact that the connecting rod, during rotation of the crankshaft, leans first to one side and then to the other. Since the piston is mounted inside the liner with a slight diametric clearance, in addition to its principal movement (which takes place in two directions, according to the axis of the cylinder) it also makes a secondary movement. It moves from one side of "rest" to the other, inside the cylinder liner. To render the movement in question more

To reduce loss due to friction, competition engines typically use only two piston rings (Kawasaki).

gradually and therefore to reduce the noise of operation, many engines are constructed with pistons in which the axis of the piston pin does not lie on the plane of the engine's centerline but is instead slightly offset from it.

The parts of the piston include the crown (the top), which acts like a true "mobile wall" for the combustion chamber; the grooves for the rings; and the skirt (the portion below the rings, which serves as a guide for the piston and the connecting rod). There are also the openings for the piston pin, which are comparatively thick since they are subject to high levels of mechanical stress.

The fact that the piston's crown is directly exposed to the combustion gases means that the upper part of this component must inevitably work at far higher temperatures than the bottom part of the piston's skirt. As a result, the crown (in which there is a greater mass of metal than farther down) expands to a far greater degree than the rest of the piston. This is taken into consideration during the design phase, and in fact when cold the lower part of the piston has a greater diameter than the upper. Indeed, determination of the correct "shape" of the skirt, on the basis of a determination of how and how much each part of the component will expand, constitutes a delicate aspect of piston design. Also of importance is the non-uniform distribution of the material from which the piston is made (particularly concentrated in the area of the opening for the piston pin). For this reason the piston is made in such a way that when cold it has a diameter much greater perpendicularly to the axis of the piston pin (along the direction in which the expansion will be less, since there is less material).

The piston receives a considerable amount of heat, which must be effectively dispersed. Until a few years ago most of this heat was transferred to the cylinder walls by the cylinder rings. More recently, due to increases in engine speeds and to the average effective pressure, the quantity of heat absorbed by pistons in a given period of time has greatly increased. To keep the crown and the area of the ring grooves from reaching overly high temperatures, oil jets are aimed at the lower part of the piston crown. This oil ensures a vigorous reduction of heat and this piston-cooling method is used today in all high-performance four-cylinder engines.

Given that the piston is a movable component subject to enormous acceleration in high-performance engines, and given that it receives (and must dispose of) a considerable quantity of heat, the most logical material to make it from is an aluminum alloy that combines reduced density (thus making the component lightweight) with good thermal conductivity. The alloys developed for this use contain a high quantity of silicon (usually from 10 to 18 percent), which is particularly advantageous since it reduces the coefficient of thermal expansion. A surface coating is often applied to the skirt to ensure a certain degree of protection in the case of deficient lubrication. (It thus provides increased resistance to seizing.) This coating is composed of thin layers of tin or zinc

■ Special piston heads with bracing and support walls have been introduced recently in engines made for the highest levels of racing (Mahle).

(with a thickness of 1 to 2 microns) or layers of resin in which particles of graphite or molybdenum disulphide are dispersed to a thickness of 10 to 20 microns. The particles are applied with sprays, immersion, or by means of serigraphic procedures, which in fact is the method most often used today.

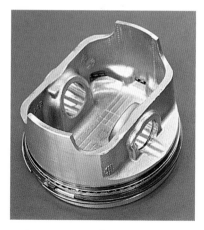

■ The great majority of pistons, like this example (made for an engine of moderate performance), are made by shell casting.

■ The inner structure of a forged piston made for an engine with high specific power

■ Forged pistons are in general use in today's high-revving engines, even though the production method costs more than casting.

Most pistons are produced by shell casting. If the intended application is particularly high stress, as in a high-performance or racing engine, then forging is the preferred production method. Forging is a more expensive procedure but one that gives the material better mechanical qualities.

In four-stroke engines, the shape of the piston crown is strongly influenced by the arrangement of the valves (or by their angle with respect to the axis of the cylinder) and by the compression ratio. To achieve a high compression ratio, the combustion chamber must have a reduced volume. If the valves are appreciably angled, it is necessary for the crown to assume a somewhat convex shape, but this is a disadvantage since it increases both the weight of the piston and the surfaces exposed to combustion gases. In modern high-performance engines with four valves per cylinder, the valves are arranged on two planes that form an angle on the order of 20 to 30 degrees. That makes it possible to create a compact combustion chamber and to use a piston

## Two-stroke Pistons

The pistons used in two-stroke engines differ notably from those used in four-stroke engines. They have a full skirt (with at most a pair of modest-sized side extensions), a more or less flat crown, and only two rings (reduced to only one in racing engines). They also have a considerable height in relation to their diameter. In these engines, which have a "square" or nearly square arrangement, the piston does the work done by valves in four-stroke engines, alternately uncovering and covering the transfer and exhaust ports. Since there are no valves in the head, there are no spaces in which to house them. Consequently, there is great freedom in defining the shape of the combustion chamber, making it possible to reach high compression ratios without giving the piston a curved top.

■ The hollow spaces designed to accommodate the valve heads are usually clearly visible on the piston crowns made for four-stroke engines.

■ Pistons for two-stroke engines are typically somewhat tall in relation to their diameter. Visible in the photo are the slots and openings at the base of the skirt.

with a nearly flat crown, even in the presence of high compression ratios.

Engines with high specific power use aggressive camshafts that give the valves increased lift and longer duration. This results in faster opening and closing of the valves as well. As a result, when the piston is near TDC at the end of its exhaust stroke, the valves are notably raised. That makes necessary recessions in the piston crown to accommodate the valve heads and to maintain a safe operating distance between components. These recessions can be somewhat deep and to a certain degree they "spoil" the shape of the combustion chamber.

In recent years, in large measure as a result of the use of computers in motorcycle engine design, pistons made for production engines have been manufactured with increasingly small skirts. This has not resulted in any problems in terms of life span or reliability. Today the skirt is often composed of two simple guide "slippers," and the height of sportbike pistons is only 55 to 65 percent of the bore.

■ Pistons for two-stroke engines usually have a flat crown with only two rings (both sealing rings).

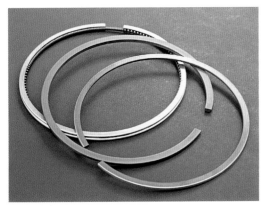

■ Piston rings are made of metal with a notch and are mounted in a groove on the piston. Piston rings seal the combustion chamber, preventing the passage of gases, and also transfer heat to the cylinder wall. Oil rings regulate excess lubricant.

## Piston Rings

Circular elements called piston, or compression, rings are used to ensure the seal between the piston and the cylinder wall. The rings are housed in grooves in the upper part of the piston. In four-stroke engines, in addition to rings—usually two for each piston in production models—there is also an oil ring that serves to remove excess lubricant from the cylinder walls. Cylinder rings have end gaps, indispensable for mounting them in the piston grooves and also to provide a necessary level of elasticity. This elasticity holds them against the walls of the cylinder. The rings are also held in place by gas pressure that expands into the hollow of the groove behind each ring, which are thus driven outward. (This is of particular importance in the case of the first ring, mounted in the highest groove.) The shape of the rings is sophisticated and requires extremely accurate production procedures. Recent years have seen a notable reduction in the thickness (height) of piston rings. There are two explanations for this. First, high-performance engines reach elevated rpm that require the use of rings with reduced mass. Second, rings with modest height are advantageous in controlling mechanical losses, which tend to grow with increases in the engine's rotational speed.

■ In four-stroke production engines there are usually two sealing rings (with an oil ring positioned beneath them). The contact surfaces are sometimes given very sophisticated coatings.

■ The pistons used in two-stroke racing engines have a single ring. As in all the pistons for two-stroke engines, there is a notch or dowel in the groove to "lock" the ring and keep it from rotating (lest one of its ends expand into a port in the cylinder).

Some oil rings have two scraping surfaces separated by a slot to return oil to the crankcase; others are made in three pieces (two steel rings with an expander spring).

Until fairly recently cylinder rings were always made of cast iron. Today steel rings (often nitrided) are in widespread use, particularly in more highly stressed engines. The work surfaces of rings are coated with thin layers of chrome, molybdenum, or chrome-ceramic.

### Connecting Rods

The connecting rod, also called the conrod, connects the piston to the crankshaft and together with the crankshaft transforms the reciprocating movement of the piston into rotating movement. Connecting rods are under tremendous mechanical stress. The small end of the connecting rod is connected to the piston by way of the piston, or wrist, pin. The length of the rod is called the beam. The big end, known as the cap, is connected to a bearing journal on the crankshaft. A connecting rod is an H-beam or I-beam according to the shape of its cross section. This shape makes the connecting rod both extremely

■ The connecting rod connects the piston to the crankshaft. The piston is held in place on the rod by the wrist pin. Visible on the piston skirt is a surface coating with a base of molybdenum disulphide.

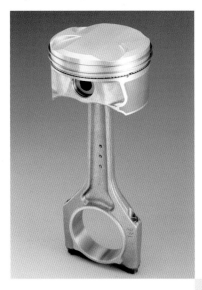

■ The connecting rods in production engines are made of high-resistance steel and are usually forged. (Only in rare cases are they sintered.) Those made for one-piece crankshafts have a head with a removable cap (Yamaha).

strong and very lightweight, indispensable qualities in such a mobile component.

The movement of the connecting rod is complex given that the small end moves together with the piston while the cap turns with the bearing journal on the crankshaft; the beam of the connecting rod thus goes through a "combined" motion. In practice, however, part of the connecting rod is said to be in alternating motion while the other is in rotation. This component is subject to traction and compression stress as well as to bending. The connecting rod also makes a pendular movement, with the piston pin acting as a fulcrum. During the engine operation, the connecting rod alternately bends to one side and then the other with respect to the axis of the cylinder, and this angle increases as the relationship between the length of the connecting rod (center distance end-cap) and the stroke of the engine decreases.

In theory, if the connecting rod had an infinite length, it would not incline at all, and the law of movement of the piston would be perfectly regular, with symmetrical movements in reference to the dead points; descending from TDC the piston would arrive at half of the stroke in 90 degrees (of rotation of the crankshaft) and would reach its maximum velocity in that position. In reality, things are different precisely because of the length of the connecting rod. The movement of the piston is determined not only by that of the crankpin on the crankshaft, which takes place parallel to the axis of the cylinder, but also by the fact that the connecting rod inclines.

The connecting rod makes two oscillations in two different directions with every turn of the crank. This pendular movement causes continuous variations in the speed of the bearing on the end of the connecting rod, even if the engine is turning at constant rpm.

In proximity to one dead point the movement is in the same direction as the movement of the crankpin, while in proximity to the other it moves in the opposite direction. The two speeds add up algebraically, and thus the speed of the bearing increases and decreases continuously.

The loads to which the connecting rod is subject result from the pressure of the gases (which explode on the crown of the piston) and from forces of inertia. These forces, which stress the connecting rod alternately with traction and compression, grow with increased engine rotation and are strongest in correspondence to the dead points, when the piston stops moving briefly before changing the direction of its movement.

To visualize what happens, think of a piston that is moving at high speed toward TDC: the connecting rod has the role of slowing it, making it stop for an instant, and then pulling it back, accelerating rapidly in the opposite direction. Thus the maximum level of traction stress is reached at TDC.

The connecting rods used in production engines are made of high-strength forged steel. The use of connecting rods machined from billet is restricted to prototypes and certain racing machines. Connecting rods made of titanium alloy are used exclusively on competition engines. Titanium alloy combines excellent mechanical qualities with a light weight. Titanium connecting rods would work very well in production engines, but the prohibitive cost rules that out, at least for now. In terms of the machining of connecting rods, special attention is paid to the cylindrical nature of the eye of the cap and the "squaring," meaning the parallels between the axis of the foot and that of the cap.

*One-piece or Split Ring*

There are two types of connecting rod, those featuring either a one-piece cap or those that are assembled, meaning made with a removable cap. One-piece connecting rods are used on built-up crankshafts and are thus used in single-cylinder and some four-stroke two-cylinders as well as all two-stroke engines. Split-ring connecting rods are used on one-piece crankshafts, always work on bushings,

■ Connecting rods made for built-up crankshafts have one-piece caps. This example, shown together with a crankpin and a needle sleeve bearing, is mounted in a two-stroke engine.

■ Traditional connecting rods are the H-beam type. The small end is called the foot, and the big end is the cap. In the example shown here, the cap is joined by means of two bolts with nuts.

■ Unlike the connecting rod to the left, this example (also made for a high-performance engine) has a cap fixed with two internal bolts and has a bearing in the foot.

and are used in most twins and all more-fractioned engines.

In connecting rods with one-piece caps, the cap works on caged roller bearings. (The exceptions, although important, are few.) The grooves in which they roll are opened directly in the eye of the cap (the external) and in the crankpin (the internal). As a result, connecting rods of this type are invariably made in case-hardened steel. The inner walls of the cap (and also those of the foot if the connecting rod is made for a two-stroke engine) are case-hardened and tempered to give them improved strength. They are also ground and lapped to give them a fine surface finish. The hardened layer usually has a thickness on the order of 0.6 to 1.9 millimeters. The remaining parts of the connecting rod are protected by case hardening (often with a thin layer of copper applied galvanically) to avoid the risk that it might become brittle.

In the case of connecting rods with split caps, the cap is joined by means of two steel or superalloy bolts with high resistance to traction. Until a few years ago these bolts fit through the rod and were retained by a nut, but then many manufacturers adopted bolts that screw directly into the rod, which does away

with the need for a nut. Most connecting rods of this type are made in hardened steel, usually alloys with such elements as nickel, chrome, or molybdenum. The resistance to traction of this material is usually between 1,050 and 1,250 N/mm$^2$.

Various methods are employed to correctly align the cap, including centering pins or sockets, or the shank of the bolts may have a calibrated center. Recently some manufacturers have begun to use connecting rods in which the cap is positioned with a sophisticated and highly accurate system of match marks, special notches cut in the two sections that are then lined up during assembly. All connecting rods made for production engines have an H-beam, which is suitable for being press forged and offers high strength with a relatively light weight. In recent years, connecting rods in which the section of the beam is an upside down H have come into use on some high-performance engines. In production engines, connecting rod length, understood as the distance between the center of the small end to the center of the big end, is typically 1.7 to 2.2 times the stroke.

In motorcycle engines the piston pin

usually floats, held in place in the piston axially by two snap rings made of steel wire. In the foot of the connecting rod the piston pin can work directly on steel, on an interference-fit bronze (or antifriction material) bushing, or on rollers in two-stroke engines.

As for the possibility of lateral movement, the connecting rods can be guided from below or above. If guided from above this is done by the cap, which is inserted with a modest axial clearance between the shoulders of the crankshaft; if guided from below, the foot is guided axially between two planes in the openings in the piston.

## Crankshaft

The cap of each connecting rod is mounted on a crankpin of the crankshaft. (Between these two components is a bearing, sliding or rolling according to the type.) The crankshaft is under great mechanical stress. It turns on main bearings and has a series of cranks, or throws. In a crankshaft, there are the main journals, which are perfectly aligned, and journals for the connecting rod, which attaches to a crankpin held between a pair of crank webs, the far ends of which often act as a counterweight and sometimes have an "axe" shape. A single-cylinder crankshaft has two main bearings and a

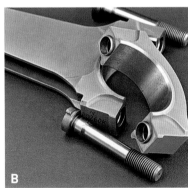
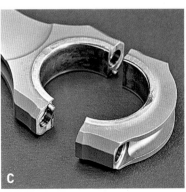
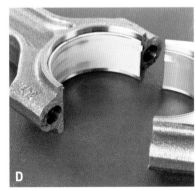

■ These four images present different methods used to obtain the correct position of the cap on a connecting rod: (A) calibrated stem, (B) calibrated bushing, (C) dowel pins, and (D) surfaces with fracture matching.

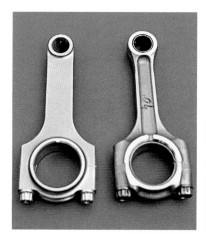

■ Here's a comparison of two connecting rods made for the same engine but using two different designs, in particular in terms of the cross section of the beam. The racing one is on the left.

■ The typical upside-down H-beam connecting rod, long popular in the field of competition engines.

single crankpin on which is mounted the connecting rod. The groups composed of the webs and counterweights have a disc shape and are sometimes called flywheels. Most of the crankshafts made for V-twin engines have a similar shape, but two connecting rods are mounted on the single crankpin, side by side. In other cases, the crankshaft has a crankpin for each connecting rod. The distance between the axes of the main journal and that of the crankpin constitutes the crank radius and is equal to half of the stroke.

Motorcycle-engine crankshafts are invariably made by forging steel, which is followed by a series of mechanical treatments. They fall into one of two categories: the built-up (or composite), which are assembled from several pieces, or one-piece. One-piece crankshafts are used in all engines with three or four cylinders and are widespread among twins as well. Built-up crankshafts are used in single-cylinders and on some V-twin engines, as well as in all two-stroke engines in which it is made necessary by the fact

that it permits the use of ball bearings both for the main bearings and for the connecting rods.

*Built-up Crankshafts*
With single-cylinder engines it is usually best, both structurally and cost-wise, to use a crankshaft built up by press-fitting three parts. In this case two shafts are used, each composed of a flywheel and main bearing unit and joined by a crankpin that is press fitted between the two units. Before the final assembly of the crankshaft, the connecting rod with its bearing (usually a needle sleeve bearing) is inserted on the crankpin. In certain four-stroke engines a sliding connecting rod bearing, composed of a large annular bushing, is used instead. The shafts are made of steel, which may be either case-hardened or hardened. (The first type is preferable in higher performance models.) The crankpin is always made of case-hardened steel, surface-hardened to a depth on the order of 0.8 to 1.0 millimeters. The holes on both the shafts and the crankpin must be given a highly polished surface finish.

The depth of crankpin insertion is generally equal to about 30 to 35 percent of the total length of the crankpin. The interference, which must be carefully measured case by case, is about 0.08 to 0.10 millimeters for a bearing with a diameter of 20 millimeters and 0.12 to 0.17 millimeters for one of 35 millimeters. After the press-fit assembly the crankshaft must be carefully "centered." Following this critical operation the two main bearings should be perfectly coaxial.

*One-piece Crankshafts*
For several years, the majority of twins and all multi-cylinder engines have used forged, one-piece crankshafts, made in alloyed steel, with a resistance to traction on the order of 1,000 N/mm$^2$

or more. (Crankshafts machined from solid billet are used only in certain racing applications.) These crankshafts turn on bushings. Only in certain two-cylinders does one find a "mixed" arrangement, with sliding main bearings and sliding connecting rods. The journals must be carefully connected to the crank-shaft shoulders (or the webs). Sharp angles and insufficient radii must be avoided since they can cause the stress that can lead to engine failure from fatigue.

Oil under pressure is fed to the bushings by way of special channels in the crankshaft. The openings through which the oil exits must have slightly rounded or flared edges. (The same applies to the entrance

■ Thanks to the connecting rods and the cranks on the crankshaft, the reciprocating movement of the pistons is transformed into the rotating movement of the crankshaft (BMW).

The simplest crankshafts are those made for single-cylinder engines. In general, they are built-up, composed of two shafts (often with discoid flywheels) joined by means of the crankpin, which is installed with a good deal of interference. Those shown here belong to a two-stroke engine (top) and a four-stroke engine (above).

openings as well, if these are in the area of the main bearings, as is the case in four-cylinders.) The journals must be worked with extreme accuracy, so as to give them the correct shape (only the smallest deviations from perfectly round are permitted), along with a highly polished surface finish. The main bearings must be precisely aligned with the highest permitted run-out error on the order of 100th of a millimeter. The main bearings on contemporary high-performance engines have a diameter that ranges between 0.44 and 0.50 times the bore. Connecting rod bearings range from 0.39 to 0.48.

In the case of four-cylinder engines,

Two crankshaft sections and the crankpin (to which will be mounted the connecting rod) that will join them. The mounting is done with a press.

the primary transmission drive gear is usually mounted as one of the flywheels and given a round shape. In the crankshafts made for other types of engines the gear is always inserted (mounted with a conical or grooved coupling). The crankshaft must be carefully counterbalanced and must possess an adequate flywheel mass. The crankshaft counterweights made for production engines are usually integral, meaning formed directly on the "extension" of the web (the part opposite the crankpin). The crankshafts in competition engines are instead given inserts made of heavy metal, usually an alloy of tungsten with a density on

This is a built-up crankshaft for a two-stroke twin engine. In the heads of the connecting rods are openings to permit the passage of oil to the roller bearings.

■ This one-piece crankshaft is made of forged hardened steel with an axe-shaped counterweight made for a V-twin. Two connecting rods would be mounted side by side on the single crankpin.

the order of two and a half times that of steel. The journals are carefully ground in two phases followed by a "superfinishing" to further improve the surface.

The metal treatments crankshafts are subjected to focus on giving the journals the necessary hardness. The typical processes used are nitriding (often the kind that uses a saline bath, which is faster and less costly) and induction tempering.

■ Crankshaft for a V four-cylinder competition engine. There are insets in heavy metal in the counterweights (Ducati).

■ This is a one-piece crankshaft, forged in high-resistance steel. One of the flywheels has been machine worked to form the drive gear for the primary transmission.

## Crankcase

The crankcase is both the support structure of the engine and the container for the crankshaft. It resembles, in terms of its appearance and function, a metal strongbox supporting and protecting the internal moving parts and sealing in oil while at the same time keeping out environmental things like water and dust. In its typical arrangement, a motorcycle crankcase is composed of two halves that are joined along a horizontal or vertical line. Above the crankcase sits the cylinder, which in turn is topped by the head. On either side of the crankcase are covers: one for the alternator and the other for the transmission/clutch unit. In many modern engines the cylinder block is incorporated in the same casting as the upper half of the crankcase. In other versions the crankcase is in two parts but has a monolithic, or "tunnel" structure, with a side or front cover. (This arrangement is used, for example, by Moto Morini and Moto Guzzi.)

*Construction Methods*

The crankcase houses the main bearings, which must be perfectly aligned. That requires machining to a notable level of precision. Single-cylinder engines and most of those with two cylinders call for a crankcase that is split vertically; the two parts of the crankcase thus have a more or less symmetrical, if not mirrored, shape. Engines with three or four cylinders feature crankcases in which the juncture of the two parts is horizontal, giving them a notably different shape from one another. The juncture line cuts the main bearings exactly in half. Crankcases of this type have a somewhat complex shape and are difficult to design and build.

Like the cylinder head and the cylinder, the crankcase is made using an aluminum alloy. Crankcases are almost invariably cast; often they are made using shell casting (using gravity or low pressure), but when possible manufacturers prefer to use die-casting, especially for models being made in

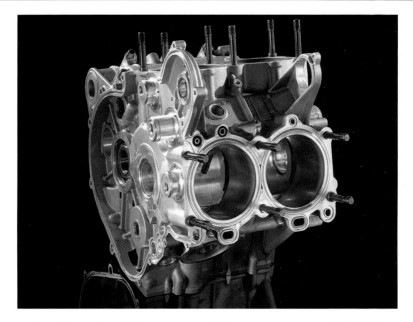

■ This sand-cast crankcase belongs to a V four-cylinder racing engine. The two-cylinder banks are open directly in the top of the crankcase (Ducati).

large production runs. Sand casting is used when only a limited number of parts is being made. The mating surfaces between the two halves and the sealing elements may be given special surface treatments. The same is true for the bearing housings and the holes for the crankcase pins or screws. To ensure the correct position of the two parts, locating pins are often used to join them.

As the lower part of the piston works it compresses or releases the air in the space beneath it. (The available volume continuously grows and diminishes during the operation of the engine.) This true "internal pumping" can lead to more than negligible losses in engines that run at very elevated speeds, causing a loss of power output. To reduce these power losses, many modern multicylinder engines feature large ventilation holes opened in the main bearings to facilitate the passage of gases in the lower areas of the adjacent cylinders.

The "pumping" in question still takes place, but the resistance created by the gases is diminished.

### Bearings

Between a shaft, turning at a certain speed and subject to certain loads and its support, there is usually a bearing. That bearing must receive adequate lubrication. The only exception, inside an engine, is the camshaft, the bearings of which can operate in housings cast directly in the light alloy of the head and the cover. In this case the loads are modest and the speed is halved compared to that of the crankshaft; furthermore, because of copious lubrication, the bearing effectively "floats" on the aluminum alloy (which itself qualifies as a more than adequate antifriction material suitable for various applications).

The simplest type of bearing is that used for sliding friction. This category includes various bushing and bearing designs. Strictly speaking, bushings, in

In single-cylinder engines the crankcase is usually composed of two symmetrical parts that are joined along a vertical centerline.

This is the crankcase half of a four-stroke single with the internal parts mounted. Note the auxiliary balance shaft in front of the crankshaft (BMW).

133

■ Half of the crankcase of a boxer twin with the crankshaft and the auxiliary shaft fully visible (BMW)

their modern version, should be called "thin-shell bearings" for they are composed of steel shells to which is applied a layer of antifriction material that must possess certain characteristics. The total thickness of the bushings used in motorcycle engines is about 1.2 to 2.0 millimeters. The layer of antifriction material in general is from 0.14 to 0.28 millimeters thick and, in the highest performance versions, is often composed of a base material and an

■ In inline four-cylinder engines the crankcase is usually composed of two parts that are joined along a horizontal line. This upper half reveals the arrangement of the main bearings (Kawasaki).

■ This view of the lower part of a crankcase made for a four-cylinder engine shows the housing for the gearbox at the back of the crankcase (Kawasaki).

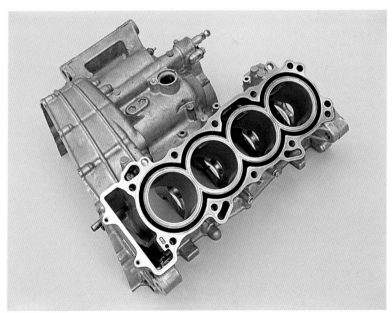

■ The crankcases of modern multicylinders often incorporate the bank of cylinders and are true masterpieces of metal casting.

■ The twins made by Moto Morni have a one-piece crankcase (a tunnel structure closed on the side by a cover) with integral cylinders and removable liners. (They are wet sleeves, with an upper flange.)

overlay that is about 0.02 millimeters thick separated by a nickel barrier.

Significant evolution has taken place recently in the field of antifriction materials because environmental regulations have led to the elimination of lead as a component.

Aluminum and tin continue to enjoy widespread use as base materials. The primary characteristics a bushing must possess include conformability (the ability to adapt to small geometric impressions in bearings and to "compensate" for slight errors in alignment, such as those that can be caused by slight flexion of the crankshaft), and absorbability (the ability to "absorb" small foreign particles that

might be present in the oil without suffering damage). Extremely important are the capacity to support loads and a resistance to fatigue. Outstanding characteristics of bushings (which function in hydrodynamic regimes and require a copious and continuous supply of oil) are their reduced radial bulk and the fact that they can be used in supports that can be disassembled and in connecting rods with caps.

A roller bearing is composed of two rings of steel (internal and external) between which are a series of ball or roller bearings. In the case of main bearings and those that support the gear shaft, the external ring is mounted in the housing with slight interference and stays firmly in place while the inner ring turns together with the bearing around which it has been inserted. Ball bearings and roller bearings are usually separated from one another with their movement controlled by cages. In connecting rod bearings the courses in which the rollers turn are not impressed in the two rings but directly in the crankpin and in the eye of the connecting rod (which are thus case-hardened and have a very high surface finish).

Roller bearings have much less need of lubrication than shell bearings.

### Sealing Elements

Sealing elements are used to keep fluids from passing between two mechanical parts after mounting or to obtain perfect airtight seals. If one part is moving (a rotating shaft or one traversing, making a linear movement) and the other is fixed, an oil ring will be used; if instead there is no movement since the parts are firmly connected to each other, gaskets are called for. In

In all two-stroke engines and most four-stroke singles the connecting rod bearings use caged rollers (Kawasaki).

certain cases there is no need for a sealing element because the function is performed by a thin film of a sealing compound (anaerobic or silicone) between the surfaces being joined.

Not all gaskets operate under

This is part of the crankcase for a two-stroke engine showing the main bearings and those for the gear shafts, all spherical.

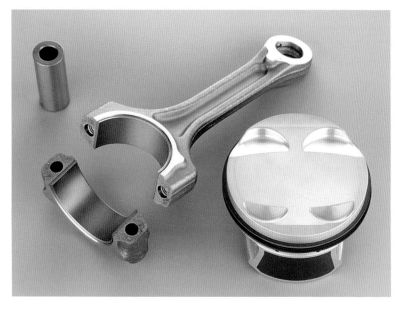

■ In all engines with a single-piece crankshaft the connecting rods work on bushings (each of which is divided in two half shells), such as that visible in this image (BMW).

identical conditions, a factor that must be taken into consideration. Of particular importance in this regard is the proper functioning of the cylinder head gasket, which must perform its function flawlessly in the presence of elevated temperatures and, especially, of sometimes powerful spikes in pressure. This gasket is designed to provide an airtight seal for the cylinders, avoiding leaks of gases, oil, or coolant.

Furthermore, it is forced to operate under difficult conditions because the pressure with which it is held between the head and the cylinder is not constant but varies with the temperature.

Additionally, it must also handle thermal expansions that cause small movements between the contact faces of the head and the cylinder.

While many of the gaskets used for the side covers are made of nothing more than chemically treated paper, the head gasket is far more complex. Today, wide use is made of MLS (Multiple Layer

Steel) with coatings in elastomer at the passage areas for the coolant channels and the oil galleries and with a film of elastic sealing compound on the upper and lower surfaces.

Simple copper gaskets or a pair of o-rings (rings made of synthetic rubber) positioned in special channels are used between the cylinders and the head of two-cylinder engines. The valve cover gasket is often made of a reusable type of rubber. (All the other gaskets are single use, meaning they must be replaced with each disassembly.)

Dynamic sealing is provided by oil seals. These are used to provide an airtight seal between a metal wall and the turning crankshaft. Typically there is one on the output shaft of the gearbox as well, located right behind the final drive sprocket. In many engines the generator works dry, thus an oil seal is used on the crankshaft, located immediately outside the main bearing on the side of the generator's rotor.

In two-stroke engines the crankcase must be airtight so there are two oil-control seals on the crankshaft, one on the side of the primary transmission and the other on the side of the generator.

Other sealing elements of this type work on shafts that are not in rotation but slide side to side. There are thus oil seals on the fork and oil seals positioned above the valve guides, which reduce to negligible levels the entry of lubricant between the valve stems and the valve guides. A typical oil seal is composed of a metal ring, usually with an L cross section, set inside a synthetic rubber body. The rubber has a sealing lip that "grips" the shaft. To increase the radial load (or the force with which the lip grips the shaft) a metal spring is almost always used.

# Fuel Delivery, Combustion, and the Environment

# INTAKE
# AND EXHAUST

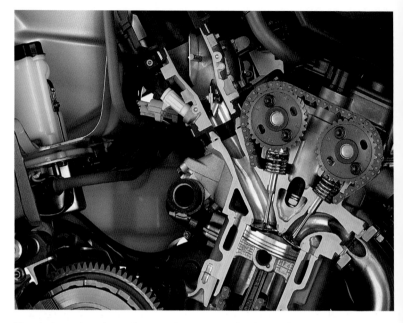

The air drawn into the engine must be free of foreign particles that could cause rapid wear of such parts as cylinder rings and cylinder liners. For this reason all motorcycles are equipped with efficient air filters,

■ Modern high-performance engines have large-diameter intake ports with an almost straight path that provides elevated volumetric delivery at high rpm (Kawasaki).

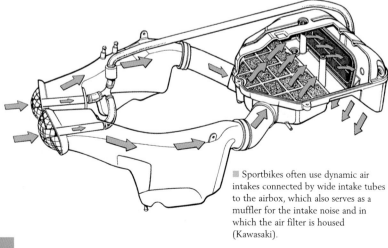

■ Sportbikes often use dynamic air intakes connected by wide intake tubes to the airbox, which also serves as a muffler for the intake noise and in which the air filter is housed (Kawasaki).

specially designed for this type of use. The particles in question are often extremely small in size (only a few thousandths of a millimeter) but still offer the threat of notable abrasive action. Unfortunately, they can be found, albeit at a lower level, in the seeming clear and clean air that one encounters on most asphalt roads. If you drive near a worksite or an area with road work in progress or if you follow a truck transporting dirt or sand, the situation obviously becomes far worse. Even more damaging is the air that off-road bikes and those used in rally raids in Africa are forced to take in.

An air filter must have both elevated filtering power, meaning it must be able to hold back even very small particles, and it must have a high accumulation capacity. At the same time, it must offer the minimum possible resistance to the passage of air so as not to penalize engine performance to a notable degree. Modern air filters often use a panel of chemically treated paper or foam impregnated with oil mounted inside a housing (the airbox) with a cover. To increase the work surfaces (meaning increase the filter's accumulation power while also pre-senting less resistance to the passage of air) without increasing the size of the filter, these elements are often made with an accordion fold.

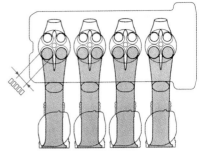

A schematic diagram from above of the intake ports on a four-cylinder sportbike. Visible is the rectilinear arrangement and the four valves per cylinder (Suzuki).

### The Airbox

The engine must take in calm air, free of turbulence, drawing it from a "lung" of considerable size. This air is drawn from the same box in which the filter is housed, and the unit as a whole is called the airbox. In most high-performance road bikes this large plastic box is located between the two upper beams of the frame, immediately above the engine. The complete system (composed of the airbox plus the air filter and the intake tubes) is designed to silence the intake of air. This is of considerable importance since the acoustic emissions of this type are far more conspicuous

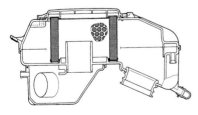

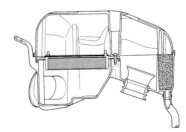

In modern four-cylinder sportbikes the intake ports are inclined sharply upward, and the large airbox is located between the frame's two side beams behind the steering head. The most common air filter is the panel type, as shown in the diagram at right (Suzuki).

■ At very high speeds, dynamic air intakes can provide a slight supercharging effect. More important is that the intakes draw in fresh air and that it flows easily to the airbox (Kawasaki).

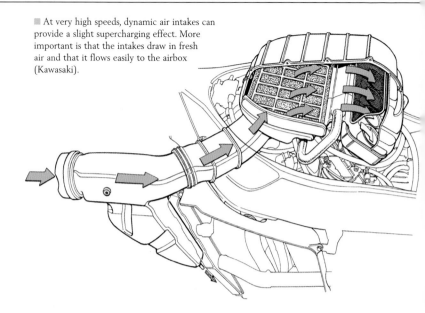

than one might imagine, and containing them without creating a measurable loss of load (meaning a significant resistance to the passage of air) is anything but easy. For this reason much use is made of Helmholtz resonators.

**Intake Tubes**

In sportbikes the airbox is fed by one or more large-diameter intake tubes that are fed by one or two dynamic air intakes located in the front of the bike, usually in the nose of the fairing, to the sides of the headlight, or immediately under or over the headlight. In this way cool, fresh air is drawn into the engine, air that has not been heated by passage through a radiator or other parts operating at high temperatures. This is significant since the air becomes more dense as its temperature diminishes. This means that, given the same amount of pressure, an equal volume of cooler air contains a greater number of gas molecules. For internal combustion engines, the focus is the amount of oxygen, since it is the constituent of

air that takes part in combustion. Thanks to special sensors, modern injection systems are usually able to take this into account. With the increase in the amount of available oxygen, the central processing unit increases the quantity of fuel emitted by the injectors. The result is that the individual power strokes that take place inside the cylinder are more powerful, and as a result, both the torque and the power output increase. Any rider can report that at increased altitudes (typically on mountain roads) engine performance declines. The reason for this is that the density of oxygen diminishes with altitude.

If well designed, dynamic air intakes can make possible, in certain situations, a kind of "free supercharging." In those conditions the air is literally blown into the airbox by the speed of the vehicle's advance; however, to achieve a perceptible result one must be going very fast. To give an idea, moving at 200 kilometers per hour you can achieve a theoretical increase in power on the order of 1.8 percent, rising to a little

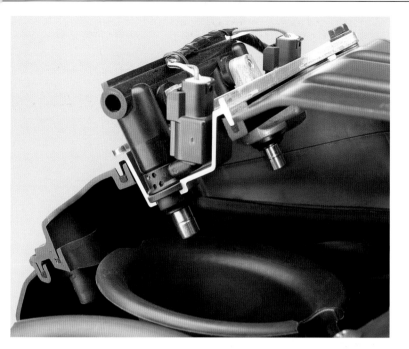

■ Sometimes the injectors are located immediately above the flared intakes, inside the airbox. In some models two injectors are used for each cylinder (Kawasaki).

under 3 percent at 250 kilometers per hour. At a speed of 150 kilometers per hour the increase is on the order of only 1 percent and at lower speeds it becomes negligible.

## The Exhaust System

Burnt gases are expelled from the cylinder with great energy and at high speed. If these gases were directed along an open tube and directly into the atmosphere, they would make a great deal of noise. For this reason motorcycles made for normal road use are always fitted with mufflers able to reduce such noise emissions.

■ A schematic side view of an older Japanese four-cylinder's mufflers (Honda)

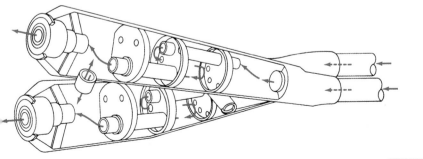

■ An exhaust system must lower acoustic emissions while presenting a minimum of resistance to the flow of the gases; further, it must permit the efficient exploitation of pressure waves to improve performance (Honda).

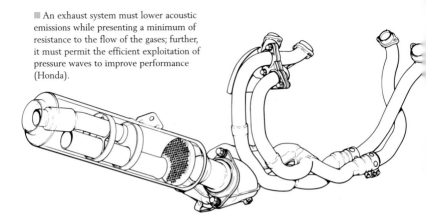

Increasingly restrictive noise laws have been put in force in recent years, and noise reduction has become a matter of extreme importance, during the design and construction of new models. A modern motorcycle's exhaust system is the result of the work of a team of specialists employing sophisticated instruments for calculation and simulation. An exhaust system passes an incredible number of hours on tests during the optimization phase of the final product.

The exhaust system must significantly lower noise emission and must do so without causing too much counter-pressure, meaning without generating resistance to the flow of the gases that would result in an unacceptable loss of power. Indeed,

■ Many mufflers reduce sound by using both reflection and absorption. In either case, it is essential that they create only a small amount of counter-pressure (Suzuki).

## Pressure Waves

Exploiting the pulsations that take place inside exhaust systems is of particular importance in the case of four-stroke engines and is also fundamental in high-performance two-stroke engines. Two-stroke engines use a system in which the gases travel down a cylindrical (or nearly) exhaust tube and then reach a divergent cone, or diffuser, at the front of a large-diameter expansion chamber that ends in a convergent cone at the end of which is a small-diameter tube called the stinger through which the gases exit into the atmosphere. The shape and size of this exhaust system are designed so that the cylinder is partially emptied by the combusted gases during the scavenge phase (when a wave of negative pressure comes through the exhaust port). The loss of fresh mixture is reduced when the exhaust port is about to be closed by the piston by the arrival of a wave of positive pressure. This happens only when the engine is running at certain rotation speeds.

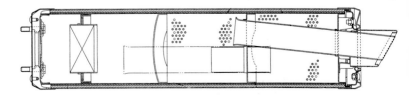

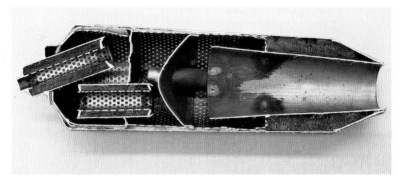

■ This photograph of the cross section of a two-stroke scooter's muffler shows the arrangement of the internal walls.

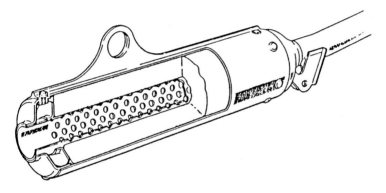

if well designed and constructed, the exhaust system can actually improve engine performance.

■ On motorcycles made for racing (those in which mufflers are used) and on motocross bikes, terminal mufflers are always of the absorption type (Yamaha).

## Exhaust Systems and Performance

Any increase in the number of cylinders increases the number of pipes that extend from the head (sometimes known as exhaust headers). These header pipes eventually gather in a collector where they are connected in some fashion. From the collector, the exhaust continues into a main pipe to the muffler. The factors at play are not only the diameter and length of the header pipe that serves each cylinder (as happens in the case of single-cylinder engines and some twins), but also the point at which the collector connection is made and how the system terminates from that point. For example, in the case of four-cylinder engines the exhaust systems can be "4 into 1" (four headers into 1 collector pipe), "4 into 2 into 1," or "4 into 1 into 2." The objective is always that of obtaining a significant diminishing of acoustic emissions while at the same time not penalizing the performance and instead contributing to obtaining the most favorable torque and power curves possible (at least for the type of use for which the motorcycle was made).

The geometry and size of the exhaust system strongly influence not only maximum performance but also the "character" of engine output.

■ In certain cases, the exhaust tubes are double walled. Sometimes aside from the main catalytic converter there is also another, smaller one located above it and called the precatalyzer (Kawasaki).

■ Chrome mufflers are usually used on cruiser bikes, often in a classical shape and sometimes with angled ends.

■ This image shows the internal structure of a modern large-volume muffler. This is the reflection type, but it may also count on the assistance of sound-absorbent material (Kawasaki).

The mufflers on sportbikes are often cylindrical and may be located at the bike's side or below or beneath the tail section (Kawasaki).

## Acoustic Emissions

Sounds are composed of fluctuations of pressure propagated in a medium. (In terms of the sounds that reach our ears, that medium is air.) In other words, they are molecular accumulations and rarefactions traveling in a sequence (forming longitudinal waves) at great speed. The speed of sound changes according to the medium through which it is moving and is strongly influenced by temperature. In the case of air, under standard conditions (1.013 bar, 15 degrees C), the speed of sound is about 340 miles per second. Each sound is characterized by both frequency and intensity. The energy transported by sound waves (sound power per unit area) is indicated as acoustic intensity. The level of this intensity is measured with phonometers and is expressed in decibels.

## Mufflers

Modern motorcycles have one or more terminal mufflers that usually have a cylindrical shape and a considerable diameter. There are basically two muffler types: reflection and absorption. Reflection mufflers have inner walls that divide the available space into chambers joined by specially arranged tubes with diminishing diameters. These mufflers are quite effective but tend to be relatively heavy and cause a greater loss of charge than absorption mufflers. Absorption mufflers have a simple structure since they use one or more inner tubes with a series of radial holes surrounded by sound-absorbent material (usually rock wool or basalt wool). In many cases special "mixed" muffler types are used that lower acoustic emissions by way of an application of both reflection and absorption.

# CARBURETORS AND INJECTION SYSTEMS

Motorcycle engines run on a mixture of air and fuel. The relationship between the quantities of the two components of this mixture is of fundamental importance in terms of the engine's correct operation, performance, and the quantity of exhaust emissions. The relationship must fall within strict parameters and well-defined limits. From a chemical point of view, the fuel/air mixture is correct when it is composed of about 14.7 parts air (by weight) for every one part of gasoline. That ratio is said to be gasoline's stoichiometric mixture (lambda = 1). During combustion, at this ratio, all the oxygen in the air combines with all the gas present, and vice versa. At the end of combustion, there are no remaining "unused" molecules of either oxygen or of the hydrocarbons that constitute the gas.

Mixtures in which the fuel is in excess of the air are said to be "rich";

■ A set of butterfly throttle valves with related injectors (visible at the base of each opening) that feed a modern inline four-cylinder engine (Kawasaki).

mixtures in which the air is in excess are called "lean." Each type of fuel has a different stoichiometric mixture. (For example, for methyl alcohol it is 6.1.) For gasoline, it is not possible to indicate absolutely precise blends since gasoline is composed of a mixture of many different types of hydrocarbons. Even so, it is on the order of 14.7.

### The Importance of the Air/Fuel Ratio

Maximum power is obtained at each degree of engine speed by feeding the engine a mixture with a rich ratio (in general around 12.5 to 13). Slightly lean blends make for lower fuel consumption. In order to get an engine to produce, overall, the minimum quantity of pollutants it must be given stoichiometric mixtures. At low engine speeds the turbulence of the aspirated mass of air/fuel mixture is quite modest, and the speed of the air passing through the intake ports is also restrained. The mixing of the fuel with the air at lower engine speeds will also be less satisfactory, especially if the fuel is being fed by a carburetor. Additionally, at low speeds, the flammability of the charge will leave much to be desired. All things considered, the engine tends to operate best if it is given a somewhat rich blend. An even richer ratio is required for starting a cold engine because during the first moments of operation, the fuel has greater difficulty vaporizing and may even form a layer adhering to the walls of the ports.

### The Carburetor
The supply of power in Otto-cycle engines is controlled by adjusting engine intake by means of a throttle valve located along the intake path above the head. The cylinders breathe freely only when the valve is completely open, meaning when the throttle grip is opened completely. The throttle setting influences the mixture ratio of air and fuel. Until fairly recently, in the case of motorcycles, it was the carburetor that provided this mixture, controlling the blend while at the same time regulating

## Simple but Complex

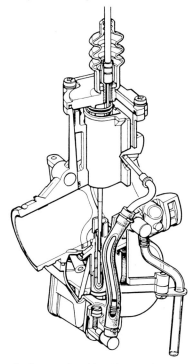

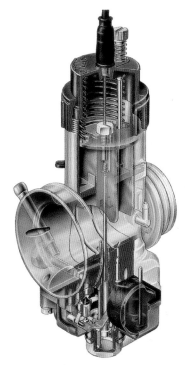

This diagram of an older carburetor shows a semiflat throttle and inclined venturi tube. Note the conical needle inserted in the fuel float bowl.

This is a transparent view of a conventional carburetor showing the shape and arrangement of the internal parts (Dell'Orto).

## Gasoline

With the exception of various dragsters and speedway cycles, the fuel used in motorcycle engines is gasoline, composed of a mixture of hydrocarbons obtained from crude petroleum through such processes as fractional distillation, cracking, alkylation, reforming, and isomerization. Hydrocarbons are organic compounds whose molecules, formed by atoms of hydrogen and carbon, can have various types of structure (straight chain, branched chain, ring, and so on). Those that constitute gas contain in general from five to ten carbon atoms.

In some countries (chief among them Brazil), where great quantities of biomass are available (composed primarily of sugar cane), alcohol is used to run Otto-cycle engines. Among the most important qualities of the various fuels are calorific power (the energy content by unit of mass) and the octane number, which indicates the antiknock power. The quantity of heat that is provided by combustion of a given volume of a mixture at its stoichiometric blend depends as much on the calorific power of the fuel as on the blend of the ratio of the mixture.

the intake by way of the throttle. Today the situation is changed, and most large-displacement motorcycles are fed by fuel injection. Even in off-road models, carburetors are giving way to injection. Carburetors continue to dominate the scene on utility vehicles and small two-wheelers like scooters and mopeds.

Throttle valve

Needle

Reservoir

Main jet

■ This cross section shows the arrangement of the elements that form the main circuit: main jet, throttle valve, reservoir, and needle (Kawasaki).

A carburetor is composed of a narrow tube (the venturi or diffuser) through which air flows toward the engine. Because of the shape of the venturi, which narrows in section and then widens again, the flow of air drawn by the engine creates a low-pressure area that draws fuel from a reservoir, or "float chamber," kept at a constant level. This fuel is passed through calibrated openings called "jets" so that the fuel is atomized and enters the tube in the form of a mist. This mist quickly mixes with the air. The quantity of fuel is regulated, and the blend is often improved by a supply of "additional" air fed by another calibrated opening.

The throttle, usually a butterfly valve, is located near the narrow section of the venture. (In effect a carburetor is a variable venturi.) To keep the blending of the fuel mixture from gradually varying from the desired ratio as the flow of air in the tube increases, there is a system of correction, which involves a conical needle connected to the throttle. This calibration element penetrates into the gas reservoir by way of the principal jet through which the measured amount of gas arrives. As the throttle opens, the needle rises and increases the available

**Minimum and
Maximum**

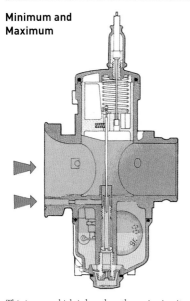

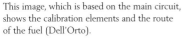

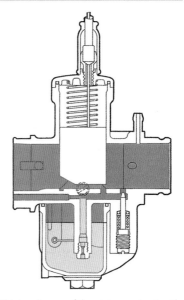

This image, which is based on the main circuit, shows the calibration elements and the route of the fuel (Dell'Orto).

This is a diagram of the minimum circuit with the related valves and jets that regulate the mixture during operation (Dell'Orto).

space for the passage of the fuel. The carburetor is completed by minimal fuel circuit and by a system of enriching for cold starting.

Constant velocity (CV) carburetors have a butterfly valve controlled by the throttle, which is located below the venturi. The conical needle is connected to a slide valve that is not operated by the driver (unlike the throttle), but that rises as a result of changes in vacuum between the intake manifold and the outside environment.

In this case, there is an intake manifold in the upper part of the carburetor (divided in two parts by a membrane) connected to the sliding valve. The greater the difference in vacuum, the more the slide, and thus the needle, rises. In this way even if the rider opens the throttle at low rpm, the sliding valve opens only progressively, performing a "moderating" function.

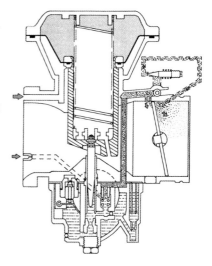

■ Here is a cross section view of a constant-velocity carburetor. The movement of the slide valve is controlled by the difference in vacuum between the manifold and the external environment (Yamaha).

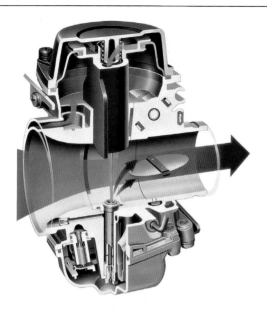

■ A transparent view of a modern CV carburetor: Visible are the butterfly throttle and the slide valve to which is fixed the needle.

## Injection Systems

When fed by the carburetor, the fuel is aspirated into the intake port. When a fuel-injection system is used, the fuel is delivered under pressure by a special electronically controlled device, the "injector." Indirect injection is used in four-stroke motorcycle engines, and direct injection is used in the few two-stroke engines that are not fitted with a carburetor. In indirect injection the fuel is fed into the intake manifold; in direct injection it is injected directly into the cylinder.

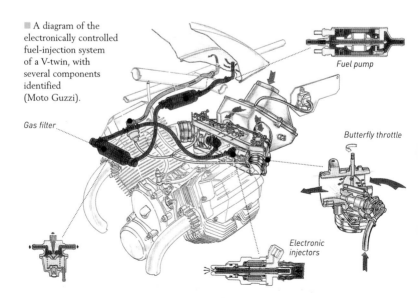

■ A diagram of the electronically controlled fuel-injection system of a V-twin, with several components identified (Moto Guzzi).

Fuel pump

Gas filter

Butterfly throttle

Electronic injectors

■ The injectors are electromagnetic and their opening is controlled by the engine's cpu.

■ Butterfly throttles with their air intakes from a modern V-twin engine.

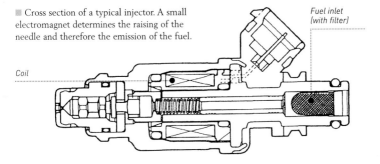

■ Cross section of a typical injector. A small electromagnet determines the raising of the needle and therefore the emission of the fuel.

Fuel inlet
(with filter)

Coil

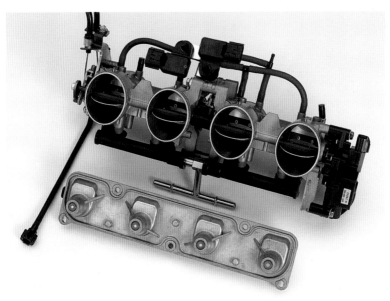

■ Some engines use slightly oval throttle bodies. A sensor informs the cpu of the opening of the butterfly throttle valves (and thus of engine load).

A modern system of indirect injection has a pump that sends the fuel under a specific pressure to the injectors (in some cases two per cylinder). The injectors are controlled electromagnetically and emit the fuel in an intermittent manner. The quantity provided at each stroke to each cylinder is modified by varying the time the

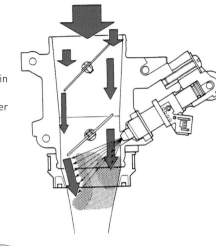

■ Some high-performance engines have two butterfly throttle valves for every port. One is controlled by the rider while the other, which has a modulating function, is controlled by the cpu by means of a servomotor (Suzuki).

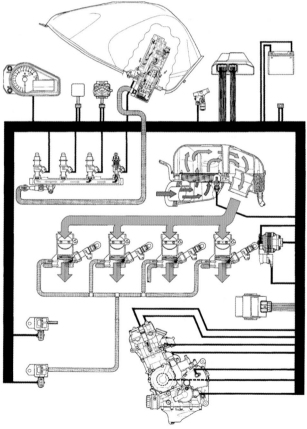

■ Modern injection systems are sophisticated. This schematic diagram shows the arrangement of the main components (the electronic unit and the sensors are clearly visible) and their locations within the layout of a four-cylinder inline engine (Suzuki).

■ This is a three-dimensional map for a cpu that controls the injection system in a high-performance engine. It makes it possible to optimize the quantity of fuel (x) furnished to cylinders on the basis of engine rpm (z) and engine load (y) (Suzuki).

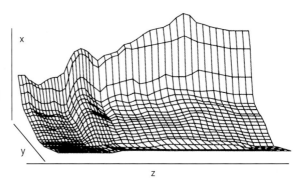

injectors are open. The control is entrusted to a central processing unit (cpu) that has been mapped and keeps track of engine rpm and the degree to which the throttle is open (or engine load). Other parameters, which allow the cpu to carry out "adjustments," are furnished by a series of sensors that

inform it about the pressure and temperature of the intake air inside the airbox and the temperature of the coolant. The unit's map is three dimensional and determines the optimum quantity of fuel to provide at each stroke to each cylinder under different conditions of operation.

■ To lower the hydrocarbon emissions (together with the fuel consumption) of the two-stroke engines that power its scooters, Aprilia has perfected an intriguing air-assisted and electronically controlled direct-injection system called Ditech (Direct Injection Technology).

# IGNITION AND COMBUSTION

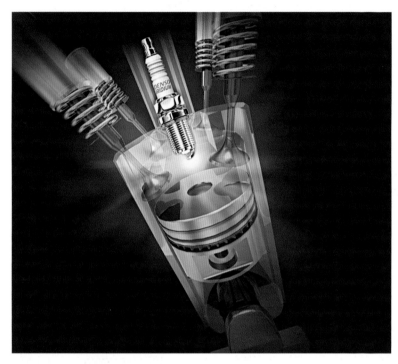

Near the end of the compression stroke, the air/fuel mixture is ignited by the spark that arcs across the gap between the spark plug's electrodes. Thus begins combustion, and rapid as it is, it still takes place over a period of time. To obtain the best results in terms of performance, the spark must flash a little before TDC to give time for the flame front to propagate and for the most advantageous pressure level to be created inside the cylinder. The best timing is not constant but varies on the basis of engine rotation speed and engine load (throttle opening).

Combustion does not begin when the spark flashes but actually a short time later. A short time after "striking the spark" the flame front begins to spread through the combustion chamber,

■ Combustion of the air/fuel mixture, during which there is development of intense heat (part of which is converted into mechanical energy), is set off by the spark that flashes between the spark plug's electrodes (Denso).

igniting the air/fuel mixture as it encounters it. This is accompanied by a progressive but rapid increase in pressure. The speed at which the flame front advances is strongly influenced by the turbulence of the combustible mixture. Careful fluid-dynamic studies and research using engine test stands make it possible to determine the optimal turbulence for every type of engine. The choice is always a compromise, for while a certain level of directed-vortex engine turbulence is needed to increase the speed of combustion, this cannot be made to happen without paying the price of a decrease in the volumetric output. To create the turbulence in question (called "swirl" if the axis of the vortex is parallel to that of the cylinder or "tumble" if it is perpendicular) causes inevitable losses of load. The speed of the advance of the flame front increases as it propagates and in fact can reach values on the order of 25 to 50 meters per second.

## Spark Plugs

The spark that ignites the air/fuel mixture flashes between the electrodes of a special device screwed into the head: the spark plug. Only its tip is resident in the combustion chamber. The threads on its body screw into a hole in the head while the rest of the spark plug is external. The position of the spark plug is critical. If it is centrally located, the distance the flame front must pass to propagate combustion to the entire mass of gas is reduced, making it possible to make use of a slight advancement in ignition. In certain cases two spark plugs are used for each cylinder (known as "double plugged"). This arrangement can be advantageous in the case of very large bores or of combustion chambers with an irregular shape.

Spark plugs operate in truly difficult conditions. The tip of the spark plug is in contact with gases whose temperature can reach 2,500 degrees C. It must also withstand very high pressure (often more than 80 bar in modern engines) that is reached in a very short time. Furthermore, spark plugs are subjected to high levels of vibration and thermal stress. In an engine turning at 6,000 rpm (a relatively slow speed for many of today's production motorcycles) the spark plug of each cylinder must

■ The spark flashes in response to a high-voltage current sent to the spark plug by the ignition system (Champion).

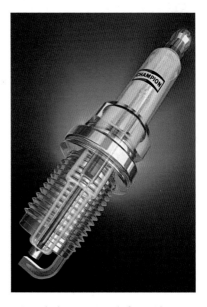

■ A spark plug is composed of an insulator (which extends the full length of the electrode) solidly joined to a metallic shell threaded so that it can be screwed into the head (Champion).

produce 50 sparks per second; at 12,000 rpm it must generate twice that number, meaning 100 per second. Even so, a spark plug must function flawlessly over thousands upon thousands of kilometers. All told, life is truly hard for this engine component which, despite its apparently simple appearance, is actually very technically advanced.

*Structure and Function*
A spark plug is composed of a metal case with a hexagonal nut (used to screw it in and out with a wrench) and a threaded shell. Sealed to the case is a ceramic insulator, and the full length of its core surrounds a central electrode. The threaded shell holds a ground electrode, and the seal between the surface of the cylinder head and the spark plug is provided by a special hollow metal washer that crushes slightly when the spark plug is screwed into the head and thus prevents any gas leakage. Part of the spark plug manufacturing process involves the application of mastic between the insulator, which is usually made of sintered aluminum, and the body of the spark plug. The mastic serves as both an adhesive and a seal. The diameter of the threaded shell varies, but the most common sizes used in motorcycle engines are 14, 12,

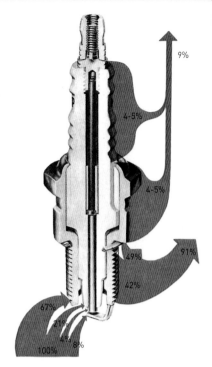

■ The arrows indicate the direction taken by heat traveling the length of the spark plug and how that heat is partitioned (Champion).

and 10 millimeters. Spark plugs also vary in length with standard lengths of 19 and 12.7 millimeters. The size of the hexagonal nut also varies. The three shell diameters noted above

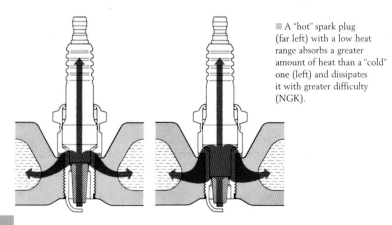

■ A "hot" spark plug (far left) with a low heat range absorbs a greater amount of heat than a "cold" one (left) and dissipates it with greater difficulty (NGK).

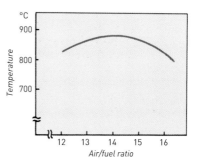

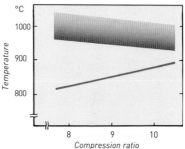

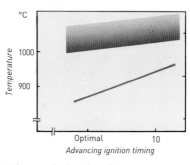

■ Use of a spark plug with a heat range that is too low can cause preignition, which can have harmful effects on the piston (NGK).

correspond to hexagons of 20.6, 18, and 16 millimeters. To function correctly, the tip of the spark plug (the insulator tip and the electrodes) must work at a temperature between about 400 and a little more than 800 degrees Celsius. Temperatures outside that range will inevitably cause problems. Temperatures that are too low will fail to burn off deposits, leading to fouling of the tip of the spark, resulting in failed ignition and misfires. Temperatures that are too high cause preignition and can also seriously damage the pistons, valves, and the spark plug itself. As is obvious, the temperature at the tip of the spark plug increases as the speed of engine rotation and engine load increase.

In order for the spark plug to work

■ These graphs show how the temperature at the working end of a spark plug (the electrodes and the tip of the insulator) are influenced by the blend of the fuel, the compression ratio, and the ignition timing advance (NGK).

■ Modern spark plugs are the result of highly advanced technology. Some use such rare metals as platinum (Champion).

correctly it must have the right heat range for the type of engine in which it is installed.

*Spark-plug Types*
Spark plugs are made not only in differing sizes with differing thermal grades but also with different electrode types. The "standard" type has a bent ground electrode that sits at a specific distance from the central electrode. This distance is called the "gap" and can vary from a minimum of 0.4 millimeters to a maximum of 1.0 millimeters.

In some spark plugs the central electrode is made with a reduced diameter, which diminishes the amount of voltage required to ignite the spark. The use of precious metals, such as platinum, can increase the life of a

## Electrodes for All Needs

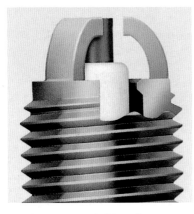

Some makers use spark plugs with more than one ground electrode, which offer high resistance to fouling and an extended life.

Increasing the distance between the electrodes improves ignition but increases the spark plug's voltage requirements.

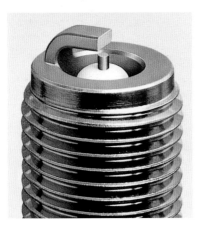

A small-diameter central electrode reduces the voltage requirement and diminishes the risk of fouling.

Spark plugs with internal ground electrodes and with very high heat ranges are made for race engines (NGK).

## Heat Range

The most important quality of a spark plug is its heat range, determined by the facility with which it dissipates the heat received from the gases and from the length of its insulator (the surfaces exposed to the gases). A spark plug that absorbs a great deal of heat (it has large surfaces exposed to the gases) is said to have a low heat range because it must then take time to dissipate the heat through contact with the head. A spark plug of this kind, suitable for low-revving engines, is said to be "hot." A "cold" spark plug has a high heat range, meaning it absorbs less heat but dissipates it easily and is thus ideal for use in a high-revving engine.

■ The use of double-plug ignition offers certain advantages in the case of large-bore engines and in those with extremely linear combustion chambers (Suzuki).

spark plug while making a central electrode of copper lowers the working temperature of the plug because of that metal's high thermal conductivity. Spark plugs with prominent electrodes are more "elastic" in terms of heat range. During recent years spark plugs with more than one ground electrode have become popular. (The arrangement is advantageous because of its resistance to fouling.) Capacitive or semicapacitive spark plugs, in which the ground electrode is composed of the same edge as the threaded shell, show up in interesting applications. Spark plugs with internal electrodes and high heat ranges offer great resistance to stress, both thermal and mechanical, and are often used in racing engines.

■ Some recent spark plugs use a central bimetal electrode (Denso). Spark plugs with prominent central electrodes, like the one on the right, are usually characterized by great "elasticity" in terms of their heat range (Champion).

# ENGINES AND POLLUTION

Controlling exhaust emissions has become increasingly important in recent years. Ever more restrictive regulations have been enacted that have led to the departure, in terms of road models, of two-stroke engines and have led designers of four-stroke engines to adopt increasingly sophisticated configurations.

Carburetors are being eliminated, particularly from models with medium to large displacement, and being replaced by fuel-injection systems. In almost all models it has been necessary to adopt catalytic mufflers. The need to lessen the environmental impact has also led to modifications in fuel composition. The gasolines of the past contained lead-based additives that made it possible to increase the antiknock power, meaning the octane number, of fuel. Unfortunately, the lead "poisoned" the catalyzer, aside from being noxious in and of itself, and it

■ Catalytic mufflers were used on automobiles well before making their way onto motorcycles. When used with diesel engines, they are combined with particulate filters (BMW).

became necessary to eliminate it from gasolines. Modern "green" fuels do not contain any lead (or only in trace amounts). It has been necessary to reformulate gases through the adoption of different types of additives.

## Pollutants

Were combustion truly complete, it would result in the formation of water and carbon dioxide; nitrogen would exit in the exhaust in the same state as it entered the cylinders, behaving as an inert gas. In reality, this does not happen, and small quantities of hydrocarbons (meaning unburnt fuel) exit the exhaust pipe along with carbon monoxide and nitrogen oxides. The admissible quantities of these pollutants, which are harmful to living things and thus have a strong environmental impact, have been progressively diminished over recent years. To achieve the increasingly severe limits imposed by antipollution regulations, manufacturers have been forced to adopt both sophisticated engine adjustments (to radically reduce emissions) and systems for the post-combustion treatment of exhaust gas.

Carbon monoxide is an extremely poisonous gas that is formed when there is a deficiency of oxygen compared to the fuel being used. Hence, the quantity of carbon monoxide produced by an engine increases with a richer blend of fuel. The hydrocarbons emitted by the exhaust are those that are not burned in the cylinders. (A small portion is also a result of the fresh air/gas mixture that escapes the cylinders during the overlap phase.) This failure to combust can be attributed to "dead" zones in the cylinder along with "interstitial

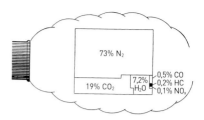

■ The quantity of pollutants is very small when compared to the total amount of gas emitted in exhaust, as indicated by this figure, which refers to a four-stroke engine (BMW).

volumes" where the flame fails to penetrate or goes out, such as inside piston rings. In this case, too, it is clear that the situation is greatly aggravated if the engine is fed a rich mixture.

Nitrogen oxides are produced in the presence of very elevated temperatures combined with an excess quantity of available oxygen. Unlike the situation with the other two pollutants, nitrogen oxides can be created in elevated quantities in mixtures that are lean and are instead negligible in rich blends.

## Advanced ignition and mixture ratio

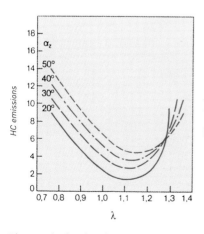

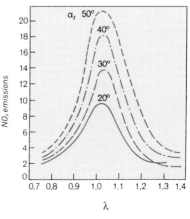

These graphs show how hydrocarbon emissions (left) and nitrogen oxide emissions (right) vary on the basis of the mixture ratio and four different advanced-ignition timings (Bosch).

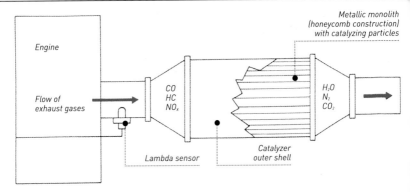

Engine

Flow of
exhaust gases

CO
HC
NO$_x$

Metallic monolith
(honeycomb construction)
with catalyzing particles

H$_2$O
N$_2$
CO$_2$

Lambda sensor

Catalyzer
outer shell

■ This diagram shows the location of a lambda sensor in a catalytic muffler that converts harmful emissions into innocuous gas substances (BMW).

## Reducing Pollutants

To reduce the creation of pollutants at the source, motorcycle manufacturers have adopted measures that involve the engine, ignition, and the fuel-injection systems. As far as engine parts go, those receiving the most attention have been the piston rings, the design of the upper part of the pistons, and the fuel delivery system. But even something as seemingly mundane as the sealing elements has been modified to improve the situation. This is particularly true in the case of the cylinder head gasket. Today, these are made of multi-layer steel, which allows a thinner gasket with stricter tolerances while also

making it possible to reduce the small gaps between the mating surfaces of the cylinder and the head.

In pistons, the height of the first ring groove has been diminished, and the sealing properties of the rings have been improved, with a consequent reduction in the loss of gases and the seepage of oil into the combustion chamber. The adoption of fuel injection systems has made it possible to obtain an improved mix of fuel and air—and to do so even when the column of gas is moving slowly in the intake ports—as well as to accurately control the blend of the combustible mixture. Electronic ignition systems with their mapping

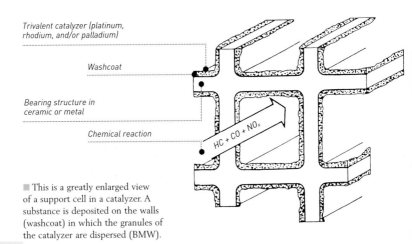

Trivalent catalyzer (platinum,
rhodium, and/or palladium)

Washcoat

Bearing structure in
ceramic or metal

Chemical reaction

HC + CO + NO$_x$

■ This is a greatly enlarged view of a support cell in a catalyzer. A substance is deposited on the walls (washcoat) in which the granules of the catalyzer are dispersed (BMW).

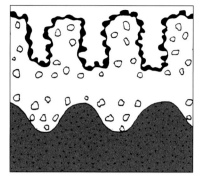

■ This greatly enlarged image shows how the washcoat, in which the granules of the catalyzer are visible, have a micro-angular surface that increases the area of contact with the gases (BMW).

controls have made a significant contribution to this. Changes in ignition timing have improved combustion quality, thereby making a significant reduction in the production of pollutants, particularly carbon monoxide.

### Catalytic Mufflers

Despite the measures taken to lower emissions at the source by reducing their production inside the engine, to keep pollution emissions within the increasingly strict limitations established by regulations it has been necessary to resort to post-combustion chamber measures, meaning those that remove pollutants from the exhaust stream before they can enter the atmosphere. An early version, used by some motorcycle manufacturers, involved the introduction of "additional" air in the exhaust port a little above the throttle so as to oxidize at least a portion of the hydrocarbons emitted. Far superior results have been obtained through the adoption of a catalytic muffler. Inside the muffler is a porous "support" (often, in the case of motorcycle engines, it is made of finely corrugated steel sheets wrapped in a certain way), with many small cells that must be crossed by the exhaust gases.

To further increase the surfaces exposed to the gases, a layer of washcoat, finely serrated, is applied to the walls of the cells. Dispersed in the

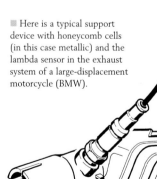

■ Here is a typical support device with honeycomb cells (in this case metallic) and the lambda sensor in the exhaust system of a large-displacement motorcycle (BMW).

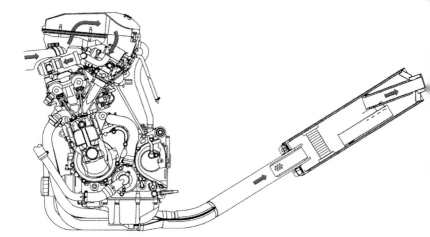

■ In this arrangement, emissions are lowered by way of a small element with catalyzers attached to the end of the muffler with a system of "additional" air drawn from inside the filter housing (Kawasaki).

washcoat are minuscule particles of catalyzers able to activate the reactions that convert the carbon monoxide and the hydrocarbons into water and carbon dioxide (oxidation) and the nitrogen oxides into oxygen and nitrogen (reduction). The catalyzers used are made of palladium, rhodium, and platinum; the washcoat is usually composed of aluminum (aluminum oxide).

### Lambda Sensors

A catalytic muffler is able to convert more than 90 percent of the harmful substances contained in exhaust gas. To function correctly, however, it must work at a high temperature; the optimal range is between 280 and 750 degrees Celsius. Excessive temperature can cause a premature "aging" of the catalyzer. It is important for the catalyzer to rapidly attain a temperature high enough for the muffler to begin its work as soon as the engine is turned on.

The most efficient conversion by a catalytic muffler is obtained when the engine is provided with a stechiometric fuel mixture. For this reason it is

### Two-stroke Engines and Pollution

Classic two-stroke engines fed by a carburetor consume elevated quantities of fuel in relation to the power supplied and also create high levels of exhaust emissions. In particular, there are notable levels of hydrocarbons, a result of poor cylinder scavenging. (A good part of the fresh mixture does not remain in the cylinder to be burned but exits the cylinder in the exhaust.) On the other hand, because of the lower temperatures reached during combustion and the fact that in general there are no excess quantities of

oxygen, the creation of nitrogen oxides is negligible.

Nothing can be done to lower the hydrocarbon emission except adoption of a direct-injection system to perform the scavenging using pure air. However, doing so is not easy and involves increased construction complexity and production costs. Furthermore, there is the fact that lubrication is by a total-loss system, meaning that some oil inevitably reaches the cylinders, where it is partially burnt, while some of it exits in the exhaust.

■ Here's the structure of a modern catalytic muffler for a high-performance motorcycle (Kawasaki).

■ To lower the emission of hydrocarbons in two-stroke engines, which are otherwise very high, some makers have developed systems of direct injection such as this Aprilia's Ditech system.

important that the air/fuel mixture ratio be regulated with great precision to keep it within the so-called blend window at all times, so that it veers only in minimal measures from the optimum mixture ratio. To achieve this end, the engine-management system employs what are called lambda sensors. Such a sensor positioned in the exhaust port informs the engine's central processing unit of the quantity of oxygen in the gases leaving the engine. The unit can use this information to determine the blend of the air/fuel mixture and then take whatever steps are necessary to bring the mix within the blend window, including diminishing the quantity of fuel emitted by the injectors. This kind of on-the-fly accuracy would never be possible using a carburetor.

169

# Transmission

# CLUTCH AND GEARS

In the typical layout of a motorcycle, the clutch and gears are part of the engine. The clutch, which is connected to the crankshaft by means of the primary transmission, is located outside the crankcase in a housing that is closed by a special side cover, while the gears are housed in the rear part of the crankcase. The clutch is mounted on one end of the input gear shaft, while the output gear shaft is connected to the sprocket of the final drive, which is most typically powered by a chain and transfers movement to the rear wheel.

There are notable exceptions to this layout. There are motorcycles that make use of an "automotive-type" transmission such as BMW's boxer twins and Moto Guzzi's bikes. In these motorcycles the clutch, mounted directly on the flywheel located at the back end of the crankshaft, has one, or at the most two, friction plates mounted on the splines of the transmission input shaft,

■ Changing gears changes the transmission ratio between the engine and the rear wheel; gears are engaged by means of sliding gears and engagement dogs that engage with freewheeling gears (BMW).

and the final drive is by shaft. The gearbox is separate from the engine, housed in a box made of light alloy that is fixed rearward to the crankcase by means of a series of bolts.

## The Primary Transmission

In motorcycles built with a conventional layout, the crankshaft's axis of rotation is transverse to the frame. Motion is transmitted to the clutch by means of a pair of meshed gears (by far the most common method used) or by means of a chain. If it is by a pair of meshed gears, there is an inversion in the direction of rotation such that the input shaft of the gears turns in the opposite direction of the crankshaft. If it is by chain, this inversion does not take place, and the two shafts turn in the same direction. The function of the drivetrain is to connect the crankshaft to the clutch, and thus the engine to the transmission. To accomplish that, the drivetrain must also ensure a certain reduction in the speed of rotation, which corresponds to a proportional increase in torque.

Primary transmissions with gears are exceptionally strong; they last longer than the engine itself and, if given adequate lubrication, require no maintenance. Since they do not suffer the effects of centrifugal forces, they are especially well suited for use in fast-revving engines. They lend themselves to better fuel efficiency, but until relatively recently it was sometimes difficult to get them to function silently, so recourse was often made to the use of gears with helicoidal teeth.

■ This image makes clear the arrangement of the parts that compose the primary transmission (by which motion is sent to the clutch) and the gears. The final drive here is by shaft (BMW).

■ The majority of today's production motorcycles use gears to drive the primary transmission. These are gears for a competition bike.

■ In production motorcycles, the primary transmission sprocket and the clutch basket are separated by a flexible coupling, often composed of several rubber inserts.

Eventually, thanks to the increased precision of metalworking and the adoption of gears with modified shapes, this problem was resolved, and for many years all primary transmissions with gears had "straight-tooth" gears, which ensure improved efficiency.

The meshed drive gear can be installed on one end of the crankshaft with a bevel gear or grooved coupling with retention nut, or it can be machined directly into one of the flywheels on the crankshaft, which will be given a round shape for this purpose. This second version is the one usually employed in four-cylinder inline engines. Primary transmissions with chains, used universally when separate transmissions were common, are today somewhat rare. They function silently and are efficient, but they are not suitable for high-revving engines and they do not have a life span comparable to those using gears. Additionally,

■ This diagram of a modern four-cylinder engine shows the arrangement of the geared drivetrain, the clutch, and the gears. The flexible coupling in this case is a spring type (Honda).

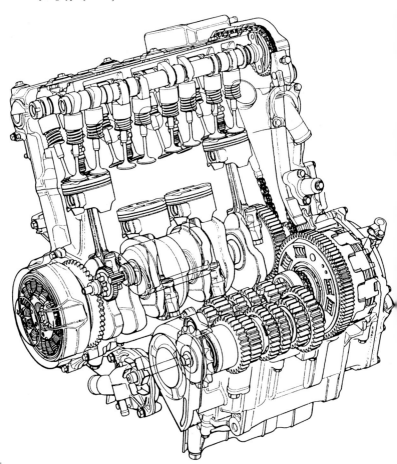

■ The most common clutch used in motorcycles is the multi-disk, working in an oil bath. This is called a wet clutch (Kawasaki).

they sometimes require a system of tensioning or guides.

Mixed primary transmissions, composed of a first part with a chain and a second with gears, enjoyed a certain level of popularity in the past. There are also two-stage primary transmissions in which both stages are gears. The first pair of meshed gears operates an auxiliary shaft that in turn moves the clutch-sprocket group, which is integral with the basket housing of the clutch.

### The Clutch

The clutch permits or interrupts transmission of motion from the engine to the gears, thus making it possible to modulate the gradual onset of engagement to ensure smooth and progressive starts. A typical motorcycle clutch is composed of a "stack" of driven plates and drive plates along with a pressure plate located inside a "basket" housing. Taken together, the clutch transfers motion through its connection to the primary drive gear sprocket. In the central, innermost area of the clutch is the driven element,

■ This illustration of a modern multi-disk clutch shows the basket, the drum, the pressure plate (driven by springs), and the "pack" of driven and drive plates (Kawasaki).

■ Shown here are the components of a competition, dry multi-disk clutch. Both the basket and the hub were machined from billet aluminum alloy.

the hub, mounted on the input gear shaft. The hub has a series of teeth, or lugs, on its outer diameter that engage teeth on the driven plates. These driven plates are made of steel and alternate with the drive plates, which are made integral with the basket housing by means of a series of teeth on their inner diameter. The drive plates are covered on both sides with a friction material. The clutch is completed by a pressure plate driven by a series of springs that press it against the stack of plates. In this way the plates themselves become integral, and motion is transmitted from the basket housing to the internal drum (and thus from the sprocket of the primary transmission to the input shaft of the gears). When the clutch lever is pulled, the springs press against the pressure plate, which compresses the plates against one another and thus disengages the clutch. In this situation, the basket turns independently of the

hub. Drive motion is no longer being transmitted from the engine to the gears and vice versa.

Along with this basic design there are variations on the theme. For example, instead of a series of springs the clutch can use a single, large-diameter central spring. Also reasonably widespread are clutches that employ an arrangement that is a "reversed" version of the one just

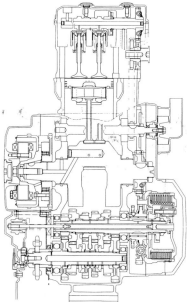

■ Clutch friction plates wear a coating of friction material on both sides and are ringed by external radial lugs (which fit into grooves on the clutch basket).

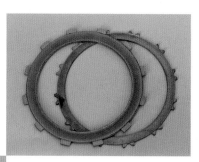

■ This diagram of a V-twin engine shows the arrangement of the two gear shafts and the multi-disk clutch. The crankshaft runs on thin bearings (bushings). To its left end is the alternator (Suzuki).

described, meaning the pressure plate is located in the innermost part of the basket (near the gears). In this case, in the exterior area there is a disengagement plate worked by the clutch lever. This disengagement plate is often connected to the clutch lever by means of a cable with one or two threaded adjusters to compensate for cable stretch.

Another common arrangement instead uses a hydraulic system not unlike the one used for brakes (fluid reservoir on the handlebar, hydraulic line, and action performed by a master cylinder in which a piston is housed). In this case, the system is self-regulating and does not require the periodic adjustment of the cable-actuated system. As with the brakes, the hydraulic fluid in the circuit must not contain any air bubbles. Unlike fluids, gases can be easily compressed, and thus any bubbles might "absorb," by compressing themselves, the variations in volume caused by the movement of the small piston in the master cylinder. The circuit would then lose its pressure, and the cylinder could not disengage the clutch.

In some more refined motorcycle engines, the gears, although installed in the back of the crankcase, can be removed without having to split the crankcase (Kawasaki).

■ This photograph of extractable gears shows the selector drum with its shaped grooves, two shifter forks, and the two shafts. The upper shaft is the input shaft that takes motion from the engine (BMW).

Flexible couplings are usually installed between the sprocket of the primary transmission and the clutch basket. These are composed of a series of rubber cushions or, in the case of motorcycles with larger displacement, of several springs. Traditional motorcycle clutches work in an oil bath since they are housed in the same space as the chain (or the pair of meshed gears) of the primary transmission. To reduce the losses from sloshing oil and to transmit elevated levels of torque without resorting to a

## Slipper Clutch

In recent years, there has been a need to limit the rear wheel's tendency to hop or shudder as a result of sudden braking or downshifting. This most often occurs in large-displacement sportbikes. Several remedies have been suggested to control this problem, usually caused by heavy braking in a curve (with the accelerator off). One of these ideas, adopted by several motorcycle makers, calls for the use of a slipper clutch, meaning a clutch made to slip when the braking torque transmitted by the engine exceeds a certain level. One solution is a mechanical system employing frontal

cams to reduce the pressure the clutch springs apply to the pressure plate at a certain point. Another system exploits the same principle adopted for the operation of automotive power brakes. This involves the use of a large pneumatic capsule attached to the intake port (below the throttle). The capsule exploits the vacuum created inside the intake valve during its inactive phase. In practice, atmospheric pressure is used to a large measure to generate a force that acts in the opposite direction of the clutch springs. This means that beyond a certain level of engine torque the clutch slips.

large-diameter clutch, some high-performance road bikes and all competition bikes use "dry" multi-disk clutches.

As previously mentioned, motorcycles with a longitudinal crankshaft (BMW boxers, Moto Guzzi), use an automotive-style clutch in which the drive element is a flywheel fixed to the end of the crankshaft, which also turns the pressure plate of the clutch. The driven element is a friction plate mounted on the splines of the transmission input shaft. The spring is a diaphragm type.

**Gearbox**

Directly below the clutch is the gearbox, which allows the rider to engage different gears in order to change the transmission ratio according to riding conditions. A lower gear will reduce the engine's rotation speed while at the same time increase available torque. Lower gearing thus makes for improved acceleration over steep terrain. High gears, on the other hand, reduce the torque available to the rear wheel while decreasing engine rotation speed.

A typical motorcycle gearbox has two parallel shafts, the input and the output (also called the main shaft and countershaft or the primary and secondary). Gears are installed on each of the shafts. The gears on the two shafts can be made to mesh, and each pair of meshed gears corresponds

Sliding gear
Fixed gear
Freewheel gear

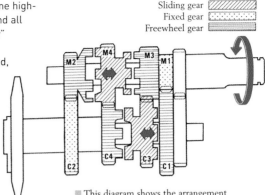

■ This diagram shows the arrangement of the gears on two gear shafts with four speeds. Those sliding (one on the primary and one on the secondary) have openings in which the forks that direct the movement are inserted (Honda).

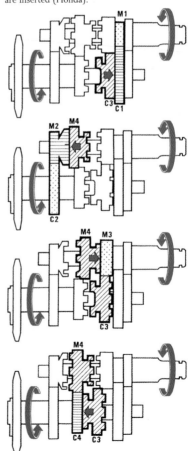

■ This diagram shows the position of the sliding gears corresponding to the four gears. From above: first, second, third, and fourth (Honda).

to a different transmission ratio. The principle of operation is simple: for every pair, one of the gears is freewheeling on its shaft while the other is engaged on its shaft. To engage the gears, the freewheeling gear must be made to turn together with its shaft. This is done by using the gears adjacent to the freewheeling gears. They can slide back and forth along the grooved length of the shaft on which they are mounted, and because of the way they are mounted on the shaft they turn together with it. On one or both sides of these sliding gears are metal teeth called "dogs." When the gears are paired (when the gear is engaged), these dogs mesh with those on the free-wheel gear, engaging it and locking the freewheeling gear to its shaft.

### Gear Selection

The gearbox contains three types of gear: fixed (solidly mounted on the shaft), sliding (mobile back and forth, but turning together with the shaft), and freewheeling. The different available gears are engaged by making the freewheeling gears lock with their shaft. The sliding gears are moved using shifter forks that move side to side along shafts on which they are mounted. The ends of the forks fit into openings in the sliding gears. As a result, the given move-ment of each fork corre-sponds to movement of the gear on which it is bearing. This axial movement is controlled by a gear drum-selector mechanism with shaped grooves into which fits the shifter fork guide pin. The drum,

■ This six-speed gearbox is housed in a light alloy case that bolts to the back of the engine crankcase. Visible are the selector drum and some of the shifter forks (BMW).

■ The selector drum is moved into predetermined positions (each of which corresponds to a gear) by means of a mechanism operated by the shift lever.

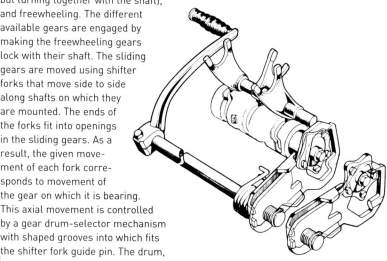

## Gears with Direct Drive

The typical motorcycle transmission uses an input shaft on the end of which is mounted the clutch, and an output shaft, on the end of which, on the opposite side, is the final drive sprocket. In the past, different arrangements were common, including those in which the axis of rotation of the final-drive sprocket and that of the clutch were not parallel but instead coincided. Three shafts were used in this system: input, output (sometimes composed of a short sleeve that was inserted on one end of the input shaft), and an auxiliary shaft, parallel to them. Also in this arrangement there were as many pairs of gears as there were transmission ratios. In all gears except one (usually fourth or fifth), motion was transmitted from the input shaft to the auxiliary shaft by gears (which varied according to the gear selected) and then from that shaft to the output shaft by means of another pair of gears (this time always the same, no matter which gear had been engaged). In one of the two highest gears, the input shaft was locked together to the output shaft, and the power was transmitted with a 1:1 gear ratio (direct drive). In transmissions of this type (which have mostly vanished from the scene with the exception of some Harley-Davidson twins and the single-cylinder Enfield produced in India), there is no inversion of motion: the sprocket of the final drive and the input shaft turn in the same direction.

in turn, is moved by a mechanism connected to the shift lever. The mechanism is designed to give a certain angular movement to the drum, which is then forced to stop in fixed positions (each of which corresponds to a gear) every time the pedal is raised or lowered.

The shifter forks move from side to side (with respect to the central one, in which the teeth of the sliding gears do not mesh with those of the adjacent gears), and therefore each of them can engage two gears. The movement of the sliding gears is limited, and each of them is always meshed with its corresponding gear on the other shaft.

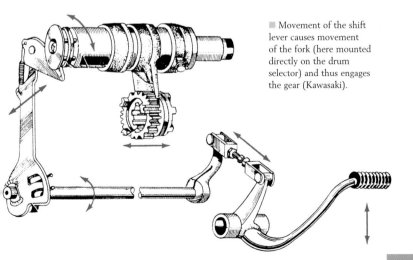

Movement of the shift lever causes movement of the fork (here mounted directly on the drum selector) and thus engages the gear (Kawasaki).

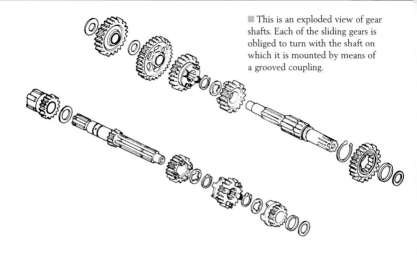

■ This is an exploded view of gear shafts. Each of the sliding gears is obliged to turn with the shaft on which it is mounted by means of a grooved coupling.

There can be slight variations to this "constant mesh" transmission system. For example, often the forks are not mounted on shafts but directly on the selector drum, the exterior surface of which is suitably machined. The two shafts of the transmission are supported by roller bearings, and since the input shaft is connected to the output by way of a pair of gears, they turn in opposite directions.

**Constantly Variable Transmissions**
CVT transmissions are completely automatic, with a stepless change of gear, and are used on all scooters. Recently they made their appearance on motorcycles. A constantly variable transmission has a trapezoidal belt that transfers movement from a "drive" pulley, mounted directly on the end of the crankshaft, to a "driven" pulley. The driven pulley is connected to the group of gears that provide the final reduction ratio.

The trapezoidal belt functions because of the frictional grip between its sides and the sides of the groove of the pulley on which it turns. The pulleys are divided in two parts, one of which is fixed while the other is movable, sliding on the shaft and making it possible to change its effective diameter. As the two parts move, the groove between them changes, expanding or narrowing and thus changing the effective diameter and radius of the belt loop around the pulleys. The belt moves upward in the groove when the radius increases and lowers in the groove when the radius diminishes. Given that the length of the belt cannot vary, any increase in the effective radius of one of the two pulleys must result in a decrease in the radius of the other.

Using this system, gears are changed by moving the mobile drive pulley (by means of a simple device using centrifugal masses); the driven pulley, on which works a spring, is obliged to do the same in the opposite direction.

Increasing the effective diameter of the drive pulley progressively alters the transmission ratio.

■ The simplicity of a constantly variable transmission, which provides a gradual change of gearing, is obvious in this photograph.

## Constantly Variable Transmission (CVT)

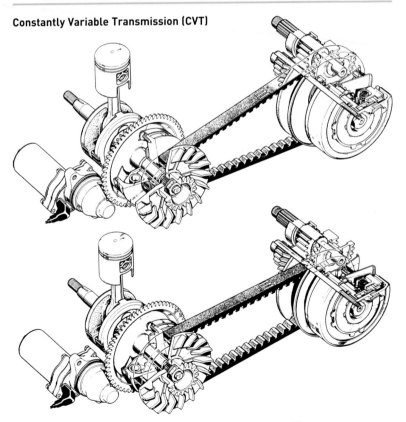

These two drawings illustrate a constantly variable transmission in two different situations (short gear/long gear). Note how the pulley width changes.

■ Motorcycles equipped with chain final drive always use roller chains. These are fitted with O-rings to make the chains more durable and to reduce maintenance (Suzuki).

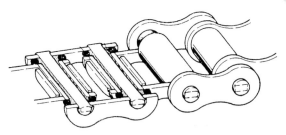

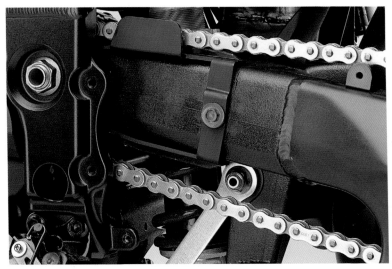

■ Chain final drive is practical, lightweight, and versatile. (It's possible to change the final drive ratio with relative ease.) They do require a degree of maintenance, however, including checking the tension and providing periodic lubrication (Kawasaki).

■ The chain links are composed of external steel side plates held together either by sleeves (known as inner links, such as those shown here), or in the case of outer links, by rivets passing through the sleeves.

## Correct Tension

Movement of the motorcycle's rear suspension will, over time, cause change in the chain tension. This is a result of the fact that in most motorcycles the front sprocket is not coaxial with the swingarm pivot. Chain tension is greatest when the axes of the wheel, the fork pivot, and the sprocket at the end of the gearbox are aligned (when they lie in a single plane). Movements away from this situation gradually cause a decrease in chain tension. Chain tension must be carefully maintained and kept adjusted as indicated by the manufacturer. Many bikes have index marks on the swingarm to aid when adjusting chain tension. Instructions for correctly setting chain tension vary with manufacturer, some requiring that the chain be checked with one wheel off the ground, others with both wheels on the ground and no one on the bike, still others with someone seated on the rear of the saddle, and so on. Such instructions should be followed precisely.

184

## Toothed Belts

In recent years, several motorcycle manufacturers have adopted final-drive systems that use a toothed belt for some of their models. The belt itself is a simple device with none of the mechanical complications of a chain. In theory, such belts are not extensible, meaning there is no need for periodic tension adjustments. (There are a few exceptions.) Furthermore, the belts work dry, eliminating the need for periodic lubrication. The belts must be replaced at a defined mileage interval following the maker's recommendation. Belts are much wider than chains, and they are not suitable for use on off-road bikes under any circumstances. Their use, today, remains somewhat limited.

### Final Drive System

In the typical motorcycle layout, motion is transferred from the gearbox to the rear wheel by means of a final drive system using a roller-type chain. The overwhelming majority of production motorcycles use this arrangement. Nearly all of these chains use O-rings, meaning small sealing rings made of synthetic rubber placed on the sleeves between the internal side plates and the external side plates. Use of O-rings makes it possible to insert lubricant between the plates when the chain is manufactured. In this way the critical coupling will be lubricated for the full life of the chain. The result is a long useful life for the chain combined with reduced maintenance. Even so, chains with O-rings require periodic lubrication with a suitable oil type (which must penetrate between the rollers and the side plates to keep the sealing elements elastic), and the tension in roller chains must be checked because they can stretch.

Chain final-drive systems are lightweight and make it possible to easily change the final drive ratio. Increasing the number of teeth in the front sprocket (or diminishing the number of teeth on the rear sprocket) lengthens or shortens the ratio. It is interesting to note that the front sprocket should never have too few

■ A motorcycle rear end featuring a single-sided swingarm and final drive via a toothed belt.

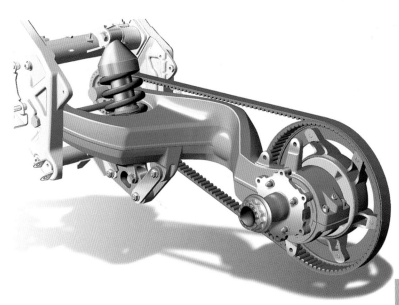

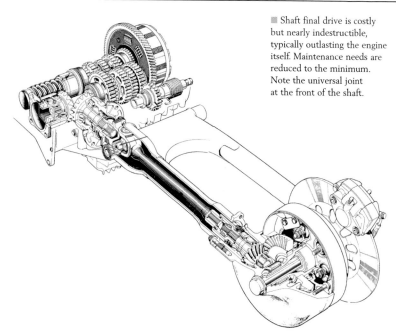

■ Shaft final drive is costly but nearly indestructible, typically outlasting the engine itself. Maintenance needs are reduced to the minimum. Note the universal joint at the front of the shaft.

teeth, which can cause irregularity in its rotation (an accentuated "polygon effect"), not to mention more rapid wear of the chain and the sprocket.

Over time and miles, chains are subject to lengthening, a result of wear around the articulations. Manufacturers provide guidance, model by model, on how much the chain can stretch before it must be adjusted or replaced. Chain replacement usually occurs along with replacement of the two sprockets.

■ This diagram shows the first version of BMW's Paralever rear suspension employing a single-sided, large-section swingarm.

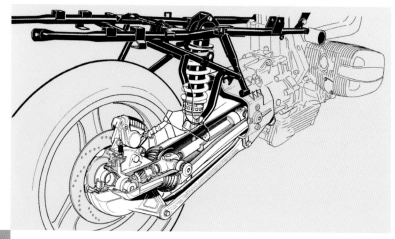

■ On its most recent transverse four-cylinder K 1200 S and R, BMW uses a reliable shaft final drive with two bevel gears (one on the transmission output shaft and one that turns the rear wheel).

## Shaft Transmission

Some large-displacement models use a final drive system with a shaft. In this case motion is transferred from the gearbox to the rear wheel by means of a shaft, usually housed inside the swingarm, with a ring-and-pinion gear mechanism to turn the axis of rotation 90 degrees. This pair of gears is located inside a light-alloy housing fixed to the end of the swingarm. The system incorporates at least one universal joint. (Very often there are two.) There is often a flexible coupling as well. Today's shaft final-drive systems are highly advanced and free of the weaknesses that characterized units of the past. In particular, they are no longer prone to such reactions as "shaft effect" (in which the bike rises and falls in reaction to throttle input), and they are no longer more "rigid" than chain final drives. On the negative side, there is still the cost and the mass of the ring-and-pinion gear and its housing. Changing the final drive ratio is complicated and costly. On the upside, modern final drive systems are very strong, extremely long lived (lasting longer than the engine), and require only minimal periodic maintenance.

A third alternative for final drive is a toothed belt (see page 185). Belts are quiet and require little maintenance, though they can be expensive to replace.

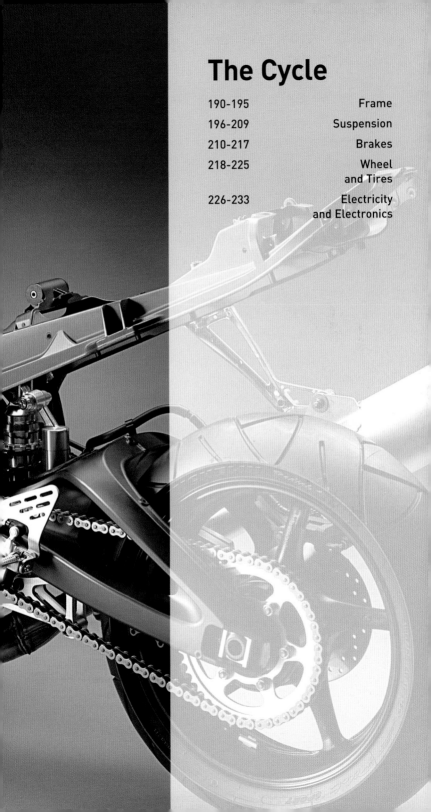

# The Cycle

# FRAME

The motorcycle's frame supports the engine/transmission components, the gas tank, the saddle, and other elements of the motorcycle's "body" and is connected to the wheels by way of the suspension. At the front of the frame is the steering head tube and at its rear is the pivot point for the swing-arm. Frame geometry is fundamental in defining the way the motorcycle handles on the road. The frame's specifications constitute the "characteristic" measure-ments of the motorcycle. The primary specifications are the wheelbase, trail, and rake (also called the head angle).

The wheelbase is the horizontal distance between the centers of the wheels and is usually measured with the motorcycle unloaded. Wheelbase is influenced by swingarm length, the steering tube angle, and front fork length. The trail is the horizontal dis-tance at ground level between a vertical passing through the axis of the front wheel and a vertical passing through the steering axis and intersecting the ground. It influences the handling and stability of the bike. Trail is influenced

■ The triangular structure of a modern tubular trellis frame is clear in this picture of a Ducati twin.

by the diameter of the wheels, the angle of the steering tube, and the offset of the front fork triple clamp (the distance between the steering axis and the plane on which lie the axes of the fork tubes), aside from the height of the bike. The rake, or steering axis angle, is usually the same as the angle of the head tube and refers to the angle between the head tube and a vertical line. It is usually measured in degrees.

Also critical to motorcycle handling are weight distribution and the cen-tralization of the masses. The latter can improve as the engine's physical size diminishes.

When the motorcycle is in operation, the suspension compresses and extends, causing variations in the motorcycle's geometry and also in the distribution of weight on its two wheels. During sudden or heavy braking, for example, the fork is compressed and the front part of the motorcycle lowers

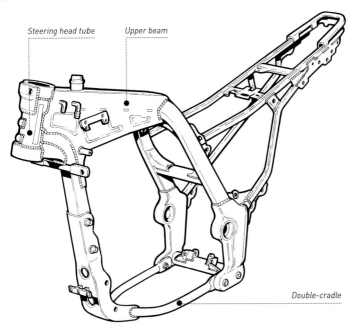

Steering head tube

Upper beam

Double-cradle

■ This double-cradle frame is made primarily of pressed steel tubes. The upper portion also serves as an oil tank as, in this case, the engine uses dry-sump lubrication (Honda).

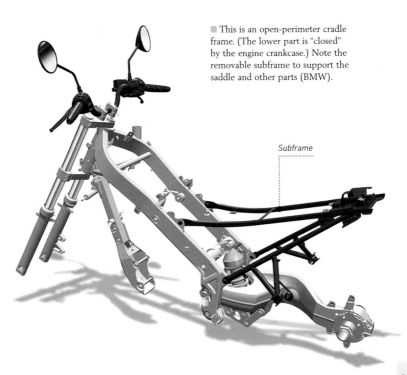

■ This is an open-perimeter cradle frame. (The lower part is "closed" by the engine crankcase.) Note the removable subframe to support the saddle and other parts (BMW).

Subframe

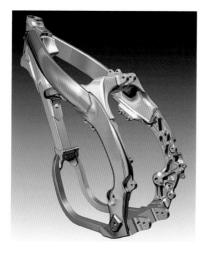

■ This split double-cradle frame is made of aluminum alloy. In this case the frame is composed of several parts made both by casting and pressing and is joined by welding (Kawasaki).

("dives"), while the rear end tends to rise. (Most of the weight transfers to the front wheel.) As a consequence, the angle of the steering head changes and the trail diminishes. Under strong acceleration the opposite occurs, ultimately culminating in the front wheel leaving the ground (a wheelie).

■ BMW's HP2 features an efficient, tubular-steel frame in a trellis structure (BMW).

## Frame Types

There are various types of frame designs. According to decisions made by the designers, not only can the overall design differ but so can the shape and composition of the individual elements that compose it. Depending on the design, the frame might be constructed of tubes, made of pressed sheet, or cast or extruded. Some of the parts may be machined from solid billet.

Tubular frames can be of either the cradle or the trellis type. In cradle frames, the part of the structure that holds the engine can be either single- or double-cradle and the lower area can be continuous or interrupted. (In the latter case, the engine crankcase serves as the lower portion of the frame.) Double-cradle frames with a single descending front tube are a common design. In this type, the frame divides at the crankcase with two tubes that pass beneath the engine and join with the rear part of the frame, immediately beneath the swingarm pivot point. The tube structure is usually completed by a triangular section in the rear that supports the saddle and is

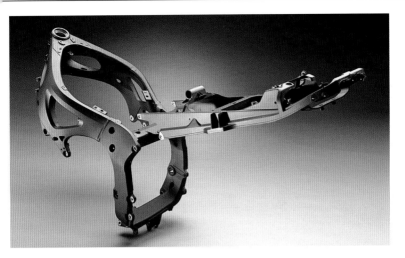

Most modern, high-performance motorcycles have an aluminum alloy frame with a twin-spar structure. This Yamaha frame shows the typical design. The subframe supporting the saddle and other parts is usually removable (Yamaha).

reinforced at crucial points by way of steel-plate gussets.

In its classic form, the upper part of a cradle frame is composed of one or two slightly bent tubes that pass beneath the fuel tank and then bend downward to reach the swingarm pivot area. In perimeter frames, the two upper elements that connect the steering head to the swingarm pivot area bend sharply downward and pass to the sides of the engine.

Trellis frames have a triangular structure composed of a large number of straight tubes welded together that work under traction or compression. (Cradle frames work under flexion and some types even under torsion.)

Trellis frames are both rigid and light. Often the engine serves an important role by completing the structure of the frame (closing the trellis) or by improving the frame's rigidity, in which case the engine is said to be a stressed member. Frames of this type are used on various high-performance road models (primarily twin-cylinder bikes,

such as those by Ducati, KTM, and Moto Morini, to which can be added the four-cylinder MV Agusta) and on some competition machines. The swingarm pivot is often inserted directly in the rear part of the crankcase. This arrangement makes it possible to reduce longitudinal bulk and puts the pivot point nearer the sprocket. Cradle frames are common on various standard models, on practically all enduro bikes, and also

Aluminum swingarms are usually composed of several parts joined by welding. (In this case only two parts were used.) This design features large sections and an upper truss for increased rigidity (Yamaha).

## Large-section perimeter frames

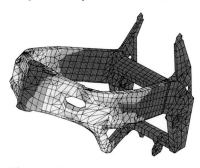

Like many other motorcycle components, frames are designed with the assistance of computer-modeling software.

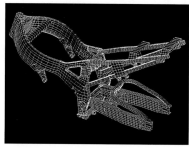

Thanks to the finite-element method of analysis, it is possible to visualize the effects of stress on structures.

on motocross and cruiser bikes. The tubes are usually round, but there are also frames made of square or rectangular tubes.

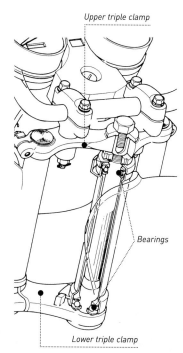

*Upper triple clamp*

*Bearings*

*Lower triple clamp*

This diagram of a steering head shows the locations of the two bearings on which the steering head turns.

Twin-spar frames offer a popular alternative use on road bikes, in particular sport bikes. In this arrangement, the steering head is connected to rear side braces in which are inserted the swingarm pivot by two somewhat inclined and robust side beams. The engine is held in place by two front downtubes and the engine also serves an essential function by contributing to the completion of the structure of the frame (which would not have adequate rigidity without it). In most cases these frames are made of aluminum alloy.

Various frame construction methods are used. In recent years, frames made entirely of cast parts welded together have enjoyed great popularity. With the adoption of advanced casting techniques, large-section parts with highly complex shapes can be made with walls of modest thickness. As a result of these production methods, the number of parts to be joined can be quite modest, which is advantageous from the cost point of view. Side beams made by joining pieces of pressed sheet steel are common. These are welded to a cast, complex front element that integrates the steering head. There are then two rear plates that are either cast or pressed and used to support the swingarm pivot. To this principal structure can be welded or bolted an additional small saddle-support frame called the subframe. In some cases, the two rear side plates are not used, and the swingarm pivots directly in the rear area of the engine crankcase, above which the two beams of the frame join. The ultimate arrangement in terms of using the engine as a stress-bearing element of the frame is the elimination of the actual frame. BMW uses this approach for some models, in which the arms of the front and rear suspension, the Telelever and Paralever systems respectively, pivot directly in the engine/gears crankcase, and the principal part of the frame is missing. In fact, there are only two small subframes that serve to support the saddle and the steering head/instrument cluster.

## Materials

The most advanced frame tubes are made of chrome-molybdenum steel. Aluminum alloy frames constructed by pressing or forging are often made in alloys of the 7000 series (aluminum-zinc), the 6000 series (aluminum-magnesium-silicon), and the 5000 series (aluminum-magnesium) in aluminum-silicon alloys (from 7 to 9 percent, in general). Lesser quantities of magnesium are used if the frame parts are to be cast. Some frames have mixed structures, composed of parts in steel joined to others in aluminum alloy, such as the large, rear plates in which the swingarm pivot is housed.

■ In some BMW models the engine/transmission unit forms the frame with front and rear auxiliary frames bolted to it.

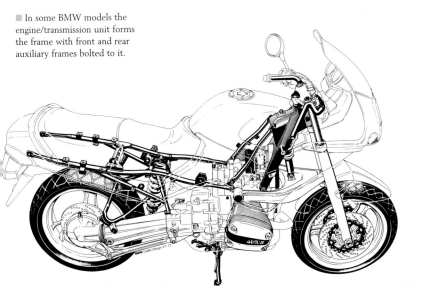

# SUSPENSION

The frame of the motorcycle is joined to the wheels by way of the suspension, which has the role of isolating the frame from irregularities in the road surface (which the wheels follow, such that vertical frame movement should be greatly reduced, if not attenuated) so as to provide comfort for the rider while at the same time giving the motorcycle the best handling at any given speed.

The suspension performs two different functions, both indispensable, that of being elastic and that of damping. The suspension system is made elastic through the use of springs, which deform under the application of force to then return to their original condition once the force has ceased. Motorcycle suspensions typically use coil springs made of wound steel-wire that, when the wheel encounters an irregularity in the road surface or when it is subjected to a force that tends to make it move vertically, compress, taking in elastic energy that the springs then release when able to return to their normal extension. The springs are not always made of steel coil. In certain cases they are pneumatic and make use of the

■ Suspensions provide an elastic connection between the frame and the wheels. Rear suspension is usually composed of a single spring/damper element, as above (Kawasaki). Front suspensions most often use a telescopic fork, as in the drawing below (BMW).

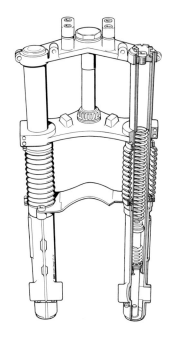

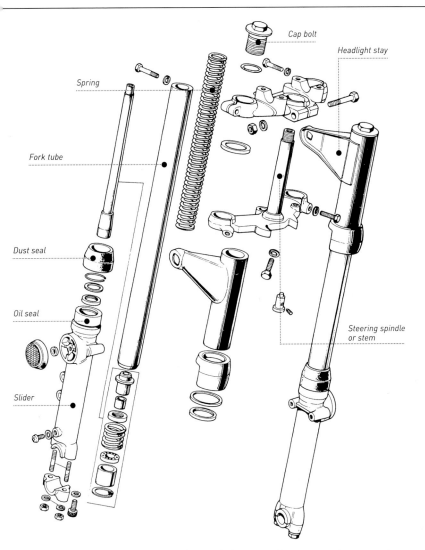

Spring

Fork tube

Dust seal

Oil seal

Slider

Cap bolt

Headlight stay

Steering spindle or stem

■ This is an exploded view of a hydraulic telescopic fork in its simplest version, without internal cartridge and without the possibility of being adjusted externally (Kawasaki).

properties of gases to diminish in volume under the action of certain forces to then expand when these forces end. Springs should be neither too rigid nor too soft, but should possess the correct elasticity for the vehicle and for the intended use.

If a motorcycle's suspension consisted of springs alone, both the motorcycle's handling and comfort would leave a great deal to be desired. In fact, once the action of the force that made them compress has ended, springs "spring back," extending beyond the length they originally had only to then compress again and then extend again, on and on through a series of oscillations of decreasing magnitude. For a motorcycle suspension this would be a "pogo-ing" sensation. To keep this

## Hydraulic Damping

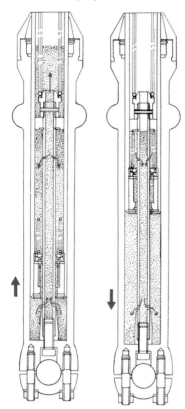

These drawings show how fork movement forces hydraulic oil to pass through calibrated holes that slow fork movement. On the left is compression; on the right is extension (Kawasaki).

from happening the suspension system includes a damping system that serves to put an end to the continuous oscillations. As we will see, however, such damping must not be excessive.

The damping system is usually hydraulic, meaning it controls the movement of the suspension by exploiting a liquid's (in this case oil) resistance to being forced to pass through calibrated holes.

### The Elastic Function

Every spring is distinguished by its "elasticity," which is measured (if the spring works through compression) by the amount of force necessary to cause a certain degree of shortening. This is indicated conventionally by the letter K. This quality, also known as the elastic constant, is expressed in newton/millimeters. The higher this is, the more rigid the spring, and vice versa.

Spring elasticity is determined by wire diameter, coil radius, the number of active coils, and material type. Spring rigidity increases with an increase in wire diameter and any decrease in coil radius and the number of active coils. The spring rate is determined by the

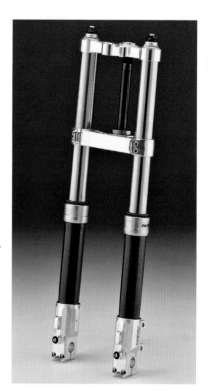

■ Motorcycle front suspensions have long been dominated by the traditional telescopic fork (shown), today largely replaced by its upside-down (USD) version.

■ USD forks provide more rigidity (the tubes are less subject to flex) than do conventional versions (Kawasaki).

distance that separates, with reference to the section, the center of the wires of two adjacent coils. The higher the rate, the greater the space between coils. The spring rate is constant if a certain load will always correspond to an equal compression. If, for example, K is equal to 20 newton, subjecting the spring to a load of 60 newtons would shorten it, through compression, 3 millimeters. Adding another 60 newtons would

■ The most advanced forks are fitted with an internal "cartridge," lending them true telescopic damping. Shown is the system of extension and compression (Kawasaki).

■ Today, USD forks (with the fork tubes below and the sliders above) have largely replaced the conventional arrangement (Kawasaki).

clamps, which in turn is attached to the steering head. The steering head is inserted in the head tube along with two bearings. The fork sliders are in the lower position, attached to the front wheel axle.

When the suspension is brought into action the wheel moves up and down and the sliders move up and down over the fork tubes. Inside each fork tube is a long, helicoidal spring that works on compression. A hydraulic unit in each slider (with a rod and piston, plus sealing elements and

shorten it a further 3 millimeters, and so on. If, however, the spring has a dual or progressive rate or is conical, or if it is made with a wire whose diameter progressively changes from one end to the other, then the spring's rigidity will vary as the spring is compressed. Some springs are made in such a way that their rigidity increases the more they are compressed. Pneumatic springs have this characteristic.

### Telescopic Forks

The most common form of motor-cycle front-wheel suspension involves a telescopic fork, which is also involved in steering the vehicle. A fork of this type is composed of two tubular components, one of which slides inside the other. The sliding member, called the fork slider, is made of aluminum; it slides over the fork tube which is most often made of steel.

In conventional designs, the slider is inserted over the fork tube with a slight diametrical clearance that lets it slide up and down. The fork tubes are at the top of the fork assembly and are attached to the two triple

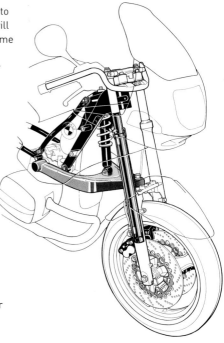

■ BMW's Telelever suspension can be seen as an intermediate type between a swingarm suspension and one with a telescopic fork.

calibrated passages) provides the damping function. The difference between the length of the telescopic fork in its position of rest (with the motorcycle unloaded but with the wheels on the ground) and its length when fully compressed is called the suspension travel. Inside each fork tube there is a specific volume of fork oil having very precise qualities. The system is completed by oil seals, perhaps bushings, and some-times also by a second spring.

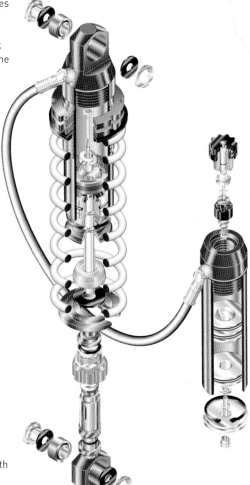

■ The internal structure of a telescopic hydraulic shock absorber with a remote reservoir containing gas under pressure, divided from the oil by means of a floating piston (Marzocchi)

## Cartridge Forks

Traditional forks provide both the spring function and the damping function at the same time, with the piston of each tube moving in direct contact with the inside surface of the fork tube. In recent years traditional forks have been almost completely replaced by so-called cartridge forks.

In these forks, inside each damping rod is a true hydraulic damper with a reduced diameter. The pistons have both calibrated holes and one-way reed valves. There are two such pistons; the upper one is fixed and mounted to the end of a long rod (fixed to the top of the fork tube), while the other is connected to the base of the cartridge's tube (meaning to the body of the damper). Cartridge forks can easily be adjusted to control the degree of damping in both extension and compression. In modern forks, the diameter of the fork tubes has been considerably enlarged to improve the flex resistance of the damping rods. The spring's rate has an influence not

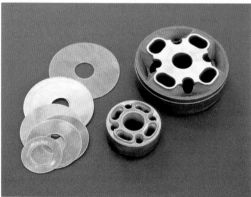

■ One end of the damper rod has the eyelet (or fork) for mounting and the other has the piston.

■ The piston has several openings. The passage of the oil takes place against the resistance offered by several overlapping elastic washers. The slowing differs according to the direction of shock movement.

only on the compression that the fork undergoes under a given load but also on the speed with which the compression takes place.

The absorption of shock is obtained by "exploiting" the resistance of a liquid (the oil in the fork, which also works as a lubricant) to being forced to pass through a restricted space. In this case, the restricted spaces are both fixed, calibrated holes and passages that are opened and closed by one-way valves. (Some of these control the hydraulic damping in compression, and others control it in rebound.) The one-way

valves are typically reed valves, composed of carefully positioned, flexible leaves. To pass, the liquid is forced to lift them. Using valves in different numbers, diameters, or thicknesses makes it possible to regulate the damping action of the fork. "Fine tuning" can be accomplished by adjusting external settings. To a certain degree, the quantity of oil inside each damper also influences its function. A great volume of liquid (incompressible) means less available space for air (compressible). When the fork shortens, the volume of available air is diminished and thus the air pressure is increased. For a given compression of the suspension, the amount of this pressure increase is greater according to how much less space is available for the air than in the state of rest. The effect of this "pneumatic assistance" increases in a large measure in the second part of the damping stroke. The viscosity of the oil contained in the fork influences the damping function as well.

### Upside-down Forks

Motorcycle suspension systems have undergone a major evolution during recent years. In terms of front forks, there has been the increasingly widespread use of upside-down (USD) forks. These invert the usual assembly structure of telescopic forks, with the outer tube fixed to the triple clamp and the inner tube positioned on the bottom. The front wheel is connected by means of "feet" located at the bottom of each inner tube. The advantages of these

■ Pneumatic springs are used in some models, usually arranged to form a single body with the damper (BMW).

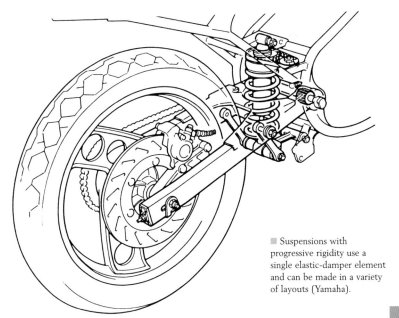

■ Suspensions with progressive rigidity use a single elastic-damper element and can be made in a variety of layouts (Yamaha).

## Spring Preload

When installing a spring or adjusting one, it can be preloaded, meaning that a certain amount of pressure can be applied to the spring while it is in its rest position. In many front forks and rear suspensions the spring preload setting can be adjusted. Modifying it changes the point at which the spring starts to respond. The more the spring is preloaded, the more force will be needed to compress it. Changing the preload, and thus changing the compression of the spring, also changes the height of the motorcycle when the rider sits on the saddle. Increasing the preload means that less compression will occur in response to a certain quantity of weight. However, the rigidity of the spring does not vary. Having passed the point of initial intervention, if the spring is cylindrical with a constant rate of coil (and if the wire has a constant diameter) a certain increase in load will result in the same compression of the spring as if the spring had not been preloaded at all (or if it had a different preload). With a more rigid spring, the same increase in load would cause less compression.

forks are that the mobile portion of the suspension is guided more accurately and that they are characterized by greater rigidity. USD forks use an internal cartridge and can be fitted easily with adjustments that make it possible to regulate the preload of the spring and both compression and rebound damping. The most sophisticated versions have large-diameter tubes with a coating of titanium nitride (TiN) on the work surfaces. This ensures exceptional sliding and great resistance to wear.

Telescopic forks function excellently. Their only drawback is that they can cause variations in chassis geometry as a function of brake "dive."

### Rear Suspension

For decades, traditional motorcycle design dictated that the rear wheel was connected to the end of a double-sided swingarm, which in turn connected to the motorcycle frame via the swingarm pivot and a shock absorber on either side of the rear wheel. This situation began to change in the late 1970s with the advent of a centrally mounted, single shock absorber. Today, the traditional twin-shock layout is rarely seen except on custom bikes or those with a retro design. Most modern bikes employ a swingarm made from aluminum alloy. A solution used by some makers (BMW and Ducati are both very fond of it) replaces the usual double-sided swingarm with a single-sided version. Another design trend has the swingarm pivoting not in the frame but instead through the rear part of the engine crankcase.

### Shock Absorbers

Movement in the rear suspension is damped by a shock absorber, typically only one, and most often located in a central position. These are quite simple in terms of their operation. One part of these is a metal cylinder (also called the body of the shock) ending in an eyelet mount. Inside the body is a piston with calibrated holes and passages controlled by calibrated valves. The piston is fixed to the end of a steel rod, the other end of which has another eyelet mount. The shock absorber is positioned between a fixed part and a moving one: the frame and the swingarm, respectively. The connection between these is often not direct but instead is entrusted to a system of links.

## Simple and Effective

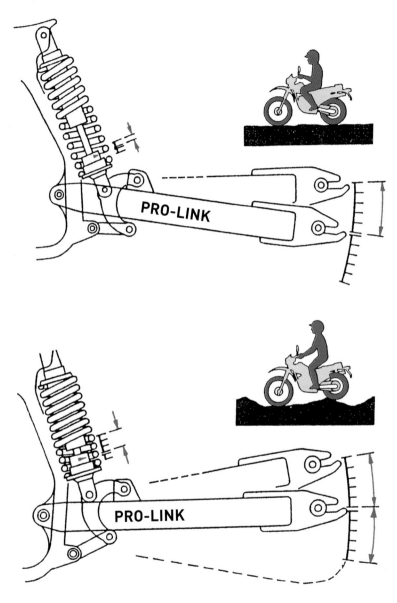

▓ These diagrams show the function of a suspension with variable rigidity (in this case the Pro-Link by Honda). Top: In the beginning, a certain wheel movement causes a small response from the spring/damper group. Above: An equal movement of the wheel after it has already passed the initial stages causes a proportionally greater movement of the spring/damper group.

During suspension movement the shock absorber undergoes variations in its length, compressing and then extending (when the distance between the wheel and the upper part of the frame increases, or when the suspension extends). The piston runs inside the cylinder, which contains oil, and it is precisely the resistance of this oil to being forced to pass through the openings of the piston that causes the hydraulic damping.

In addition to these openings there are often other fixed passages that the oil is forced through when driven by the piston. As the damper is compressed, the rod penetrates farther into the cylinder, coming to occupy a certain amount of space. Liquids are incompressible and thus if the body of the damper was

## Telelever

For several years BMW has used a special front suspension on many of its motorcycle models. This system combines some aspects of the telescopic fork with certain properties of swingarm systems. Known as the Telelever, it has a telescopic fork that in this case has only guidance functions, with the two fork tubes connected to an upper transversal element. This is connected by way of a ball-and-socket joint to a triangular swingarm, pivoting on the front of the engine, on which acts a single shock absorber. With its special design, this system makes it possible to reduce and, when needed, even eliminate the so-called brake dive common to telescopic-fork suspensions.

■ In suspensions with variable rigidity, the progressiveness of the shock absorber action is modified during wheel movement following a change in the shape of the linkage system (Suzuki).

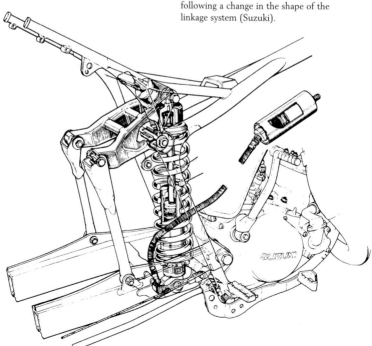

■ The central position of the rear shock and the linkages to the swingarm are clearly visible in this photograph.

■ The swingarm is sometimes given a particular shape to facilitate the passage of the exhaust can (here removed), which is often a large diameter (Kawasaki).

In this motocross bike, the upper end of the shock (note the remote gas reservoir) is bolted directly to the frame. The lower end is connected to the swingarm by means of a linkage and two connecting rods (Kawasaki).

completely full of oil when in the position of extension, there would be no space to accept the rod during compression.

It is thus also necessary that in addition to the oil there must be gas in the damper (the easiest and most obvious choice is air), for it is compressible and thus can easily compensate for variations in the internal volume. The gas must not mix with the oil since the presence of air bubbles in the oil would compromise the damper function.

There is a variety of shock absorber designs. In the twin-tube type there is a space between the internal elements (where the piston

This drawing shows the functional shape and arrangement of the elements that constitute a Kawasaki Uni-Trak suspension.

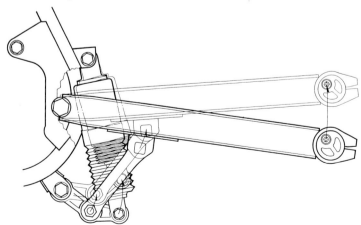

## Variable Rigidity Suspensions

Most modern motorcycles use rear suspensions that have a single shock absorber. The compression movement varies in response to the vertical motion of the wheel. If the wheel's first 30 millimeters of movement result in a certain degree of compression, the second 30 millimeters cause another larger compression. In this way the suspension is "sensitive" to small irregularities, but then gradually hardens as the vertical movement of the wheel increases. The different versions of this type of suspension vary according to how the various elements are arranged and connected (damper, connecting rod, link), but the various versions do not differ much from one another at the levels of functionality and performance. Makers often give them unique names, such as Monoshock, Monocross, Uni-Trak, Pro-lever, Pro-link, and Full-floater.

moves) and the external, which forms the body of the shock. The two "tubes" are connected by way of holes in the lower part that permit the passage of the oil. Above the liquid, in the upper area of the space, is the air, which makes possible the variations in the volume, but it has nothing to do with the oil in its rapid movement in the work area of the piston.

In the De Carbon system, the oil is pressurized by gas and is kept separated from the gas by means of a floating piston or a piston free to make only axial movements. The variations in volume can take place freely, but the gas (often nitrogen) is not in contact with the liquid. This type of damper can be considered the inspiration for those that have a remote reservoir, connected to the cylinder via a hydraulic line and containing both oil and gas under pressure, separated by a strong membrane or a floating piston.

# BRAKES

A wheel brake, at its simplest, is composed of a part that turns with the wheel and a part that is fixed to the front fork or to the rear swingarm. Although fixed, this second part is able to follow the wheel in its vertical movement (determined by the travel of the suspension). In motorcycles, as with automobiles, disk brakes are the most common system. These combine excellent performance with great resistance to wear—all at a reasonable weight. A brake of this type consists of a disk (also called a "rotor") connected to the wheel and a caliper. The caliper is the fixed element, and—whether operated by a hand lever or foot pedal—it forcefully presses brake pads situated within the caliper against each side of the disk, slowing or stopping its

■ Carbon disks are used on Grand Prix motorcycles. They are both costly and lightweight.

rotation. This caliper is controlled by a hydraulic system composed of a pump (the "master cylinder," on which the lever or pedal acts directly), and a system of brake lines terminating at the caliper. When the system is put under pressure by the rider (who pulls the lever or presses down on the pedal), pistons in the caliper press the brake pads against the disk.

The number of pistons in a brake caliper varies according to the model but ranges from one to six. There are basically two caliper arrangements, floating and fixed. Floating calipers

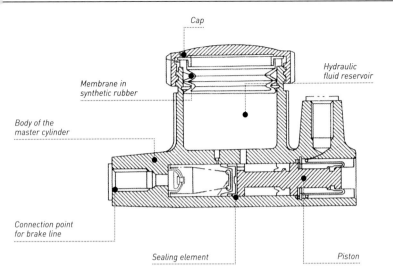

Cap

Hydraulic
fluid reservoir

*Membrane in
synthetic rubber*

*Body of the
master cylinder*

*Connection point
for brake line*

*Sealing element*

Piston

move in and out relative to the disk
and have only one or two pistons. Fixed
calipers stay in place and are arranged
on opposing sides of the disk. Some
high-performance motorcycles have
two or more pairs of pistons.

**The Hydraulic Circuit**
Brake pads must press against the disk
with considerable force, which means
that the force applied by the rider must
be greatly increased. For this reason
a hydraulic system is used. The initial
multiplication of force is obtained
mechanically thanks to the brake lever
(or pedal in the case of the rear brake),

■ This is a side view of a master
cylinder showing its internal structure
(Kawasaki).

but the second multiplication exploits
Pascal's principle: externally applied
pressure on a confined fluid is transmitted uniformly in all directions.
This is the same principle at work in
the operation of hydraulic presses and
jacks. The lever activates the master
cylinder's piston, and it, acting on the
hydraulic liquid, puts the system under
pressure. A hydraulic brake line exits
the master cylinder and connects to the

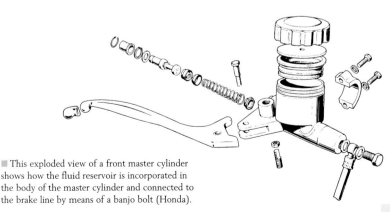

■ This exploded view of a front master cylinder
shows how the fluid reservoir is incorporated in
the body of the master cylinder and connected to
the brake line by means of a banjo bolt (Honda).

caliper (or a "splitter" if the front master cylinder feeds two calipers). The caliper pistons' diameter is greater than that of the master cylinder's piston diameter. This difference between the piston-surface diameters causes the multiplication of the force. Inside the caliper, the same pressure acts on a surface (that of the pistons) that is far greater than that of the piston in the master cylinder. Since the force is simply the pressure multiplied by the area on which it is acting, the desired result is obtained.

The front master cylinder can have an axial or radial piston (in relation to the axis of the handlebars). As a result of brake pad wear, the caliper pistons extend farther from the caliper over time, thus increasing the volume of the system. To compensate for this variation, the master cylinder is attached to a reservoir containing brake fluid that is added to the fluid already in the circuit when the pistons return to their rest position. In many cases this reservoir is integral with the master cylinder, which is usually made of aluminum alloy. For the system to function properly, there must be no air bubbles in the brake fluid. Gases can be easily compressed, unlike liquids, which are nearly incompressible.

## Fixed Calipers

If the calipers can make side-to-side movement perpendicular to the disk, they are said to be floating. If they are solidly connected to the fork or the swingarm they are called fixed. Fixed calipers use pairs of opposing pistons that firmly clamp the pads against the disk when the brake is activated. Fixed calipers can have two, four, or in some cases six pistons. The caliper is usually made of two symmetrical parts joined by bolts. There are also versions that make use of monoblock pistons. The attachment of the caliper to the fork or swingarm can be done with "tabs" and transversal bolts (with respect to the disk) or it can be done with radially arranged bolts (parallel to the disk).

The fixed calipers used most have four opposing pistons. Often the pair located on the inside of the disk has a slightly smaller diameter than the

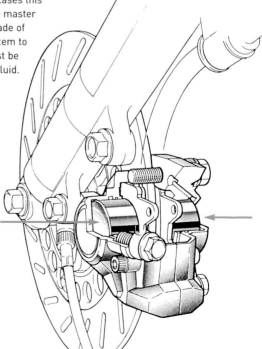

■ This is a fixed caliper with two opposing pistons. The arrows indicate the movement of the pistons when pressure is applied to the hydraulic system (Yamaha).

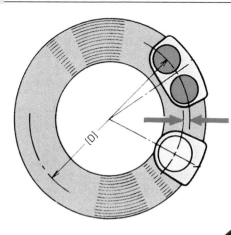

■ As indicated in this drawing, a caliper with four pistons instead of two applies a larger effective diameter (D) of braking surface to the disk (Honda).

other two pistons. This arrangement improves pressure distribution and ensures more uniform brake pad wear. In general there are two pads, but in some cases four are used, meaning one for each piston, so as to double the number of surfaces of attachment and to minimize the risk

■ This caliper with radial attachment for a high-performance motorcycle is composed of two parts bolted together (Brembo).

■ This photograph presents the two parts that compose a caliper with six opposing pistons and the two brake pads with which it is fitted (Suzuki).

## Differentiated Diameters

This half caliper, machined from solid billet, clearly shows the piston shape (in this case made of an anodized aluminum alloy). The half made for use on the inner side of the disk has a smaller diameter in order to make the distribution of pressure more uniform and thus also to make the wear on the disk pads more uniform (Discacciati).

of deformation (which would result in the uneven distribution of pressure on the working surfaces). In such cases, all pistons are of the same diameter.

### Floating Calipers

Floating calipers are not rigidly fixed to an element of the suspension. Instead, they are mounted on a support with a guide (in general calibrated pins or bushings) on which they can make slight sideways movements. The simplest version uses a single piston, but in many cases two parallel pistons are used. The body of the piston is usually in a single piece. The piston,

with its mobile pad, is positioned on the outer side of the disk, while the fixed pad is located on the inside (meaning toward the wheel) and is therefore attached to the body of the caliper. When the hydraulic system is put under pressure, the piston is driven outward and comes to press, by way of its pad, against the disk. At the same time, since the pressure in the liquid is acting in all directions, the body of the caliper runs slightly in the opposite direction, and as a result the fixed pad is also pressed against the disk, which is thus clamped between the two pads.

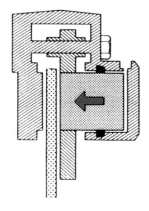

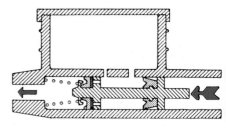

■ Side views of a master cylinder (above) and a floating piston (left). This is a single floating piston. The pressure of the hydraulic fluid pushes the piston in one direction and the body of the caliper in the other (Kawasaki) at the same time.

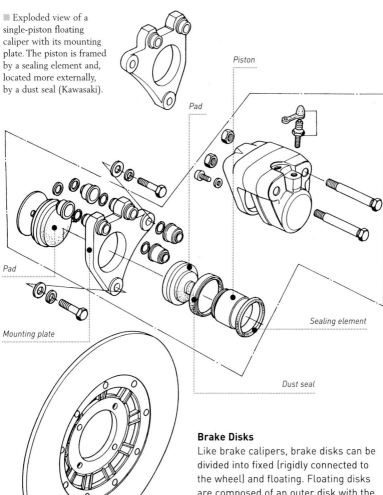

■ Exploded view of a single-piston floating caliper with its mounting plate. The piston is framed by a sealing element and, located more externally, by a dust seal (Kawasaki).

Piston

Pad

Pad

Mounting plate

Sealing element

Dust seal

## Comparison of Calipers

Floating calipers are somewhat less expensive than fixed calipers and create considerably less bulk on the side of the wheel, an important consideration when using spoked wheels, which require a certain camber. In terms of performance, however, fixed calipers seem superior to floating.

## Brake Disks

Like brake calipers, brake disks can be divided into fixed (rigidly connected to the wheel) and floating. Floating disks are composed of an outer disk with the braking surface that is connected to a disk carrier, which in turn is attached to the wheel hub by means of a series of rivets that have some play. This play allows the brake disk to "float," meaning to make a slight amount of side-to-side movement. In this way the disk, reduced to its braking surface, is freer to expand (as a result of heat) than is a fixed disk. As a floating disk expands it will change shape, automatically adapting itself to the best position between the pads of the caliper. In certain cases the disk is not connected to the wheel hub but rather to the spokes (such is the case on some

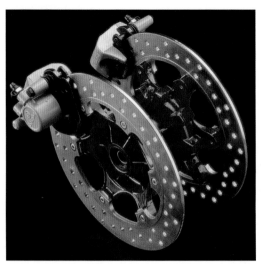

■ Floating calipers are simple, light, and compact. Their reduced bulk on the inner side facilitates the use of spoked wheels, which require a certain camber (Aprilia).

■ A floating disk is composed of a braking surface attached to a disk carrier that is bolted to the wheel hub (Brembo).

BMWs). Some contemporary motorcycles have "wave" disks that have an undulating external margin instead of a circular one. Disks often have axial holes that not only diminish weight but also facilitate cooling, "break" the veil of water when riding in the wet, and in a certain measure scuff the pads. Disks are nearly always made of steel. Cast-iron disks, though once fairly common, have largely disappeared from the world of motorcycles, although they offered excellent performance. Compared to steel disks, however, they were prone to rust and of course they weighed

■ This monoblock radial caliper has four opposing pistons. The body is a single piece, which is advantageous in terms of rigidity and light weight but makes the manufacturing process more complicated (Brembo).

## Drum Brakes

Drum brakes dominated the motorcycle scene for several decades before giving way to disk brakes, which are far more efficient. Today, drum brakes are used on some scooters and utility bikes, where their use is typically restricted to the rear wheel. They also see some use, for aesthetic reasons, on custom motorcycles. The rotating element in drum brakes is the drum, which is incorporated in the wheel hub. The drum has a cast-iron brake band, and the fixed component is a plate on which two pivoting shoes are mounted. The external, working surfaces of these shoes is lined with a friction material. The brakes are actuated mechanically with either a flexible cable or shaft. When the brakes are activated, a cam rotates and pushes the brake shoes outward to press against the drum's internal surface (or against the cast-iron brake surface). For this reason, brakes of this type are also called expansion brakes. The brake shoes are returned to their rest position by a return spring.

more. Steel disks are far less likely to incur damage, and they can be made thinner without creating problems. Grand Prix motorcycles use carbon disks, which are extremely light and offer excellent performance. Carbon is not used on production motorcycles due to the high cost and their high operating temperatures.

### Brake Pads

Brakes function due to the friction that is created between the moving part (the disk) and the elements that press against it (the pads) when the brake lever is squeezed. Each brake pad is composed of a steel plate to which is fixed a thick layer of friction material, in many cases sinterized, meaning made from a powder.

The composition of this friction material is critical in terms of brake performance (braking power, degree of modulation), pad life (in truth, both the pads themselves and the disk on which they operate), and the noise level of braking. Braking performance in rain is also determined by the type of friction material. The pad is composed of various substances, including abrasives, lubricants, fibers, metallic particles,

■ Although rare on today's motorcycles, drum brakes were once used even on competition models.

binders, "charges," and elastomers.

Motorcycle brake lines are manufactured to rigorous specifications and are made in "rubber" (the external sheath in elastomer, the middle layer in rayon or another tough fiber, and the inner layer in elastomer) or with an external sheath in braided steel or aramadic-resin fibers. The center hole has a diameter on the order of 3 millimeters.

# WHEELS AND TIRES

■ Modern high-performance motorcycles have wheels cast in aluminum alloy (Kawasaki).

Although they perform the same function as on an automobile, motor-cycle wheels are somewhat different from those used on four-wheelers, and not only because of their different sizes and shapes. In the conventional version (the sole exception being the cantilevered wheels used on single-sided swingarm suspensions), motor-cycle wheels are installed on an axle supported at both ends and have two wheel bearings in the hub. The axle is inserted through these bearings, and the hub turns around them.

Motorcycle wheels are divided into two large groups: those with spokes and those that are cast in a single piece. Spoked, or wire, wheels, which can be seen as distant relatives of the wheels used on bicycles, are used on motocross and enduro motorcycles, on some standards, and on a fair

number of custom bikes. A typical spoked wheel is composed of three parts: the hub, in which are mounted the axle bearings and the inner ends of the spokes and that usually includes one or two disk brakes; the rim, on which the tire is installed; and the spokes, which tie the hub and rim together. The traditional arrangement calls for the spokes to be installed with the heads in the hub and with the threaded ends (onto which the nipple is screwed) on the rim. The spokes, which are brought into tension by tightening the nipples after mounting,

## Ideal for off-road

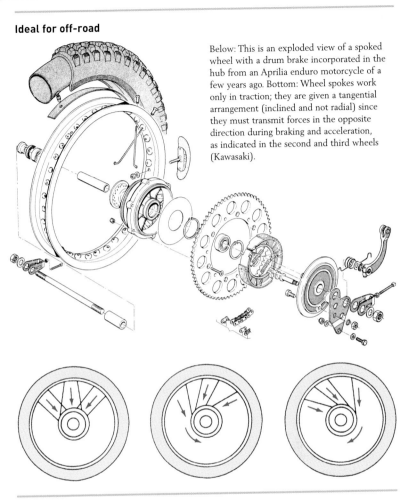

Below: This is an exploded view of a spoked wheel with a drum brake incorporated in the hub from an Aprilia enduro motorcycle of a few years ago. Bottom: Wheel spokes work only in traction; they are given a tangential arrangement (inclined and not radial) since they must transmit forces in the opposite direction during braking and acceleration, as indicated in the second and third wheels (Kawasaki).

should be given a tangential arrangement to the hub, not a radial arrangement. In other words, when observing the wheel from the side, they should seem to lean and are not arranged geometrically like the radii of a circle. They work exclusively in traction, and thus only the upper spokes, moving spoke to spoke as the wheel turns, support the load of the motorcycle. The tangential arrangement is explained by the fact that during the operation of the motorcycle the spokes must also support the considerable forces that

are generated in braking and acceleration (which would not be possible if they were perfectly radial). Because they are inclined alternately in both directions, the spokes crisscross one another. To provide the wheel with adequate rigidity also in the transverse sense, the spokes must be arranged so as to have a certain inclination when the wheel is viewed from the front of the bike. This is the cambering of the wheel. With reference to the transverse section of the wheel, the classical arrangement calls for the spokes to be fixed to both

■ The bearings on which wheels turn are housed in wheel hubs. After the axle nuts are tightened, the bearings' inner races are firmly attached to the axle and thus are fixed; the outer races, instead, are joined to the wheel (Kawasaki).

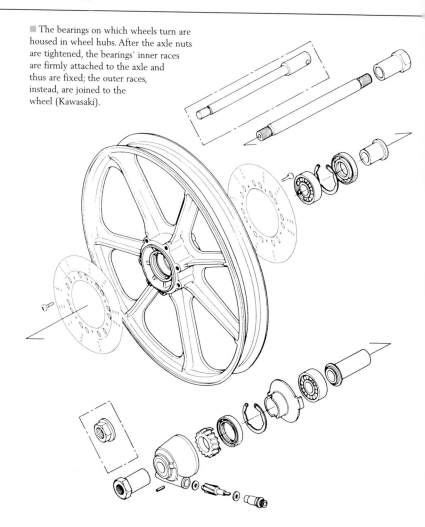

sides of the hub and to the central area of the rim. Mounting is done by putting the spokes in their correct working position and then tightening the nipples to firmly join the rim to the hub. After mounting, the wheel must be balanced. At the end of that operation, the wheel should turn without wobbling, both in the radial and the transverse sense. In other words, the plane of the center line of the wheel will be perpendicular to the axis of rotation, and the wheel will be perfectly coaxial to the hub.

Spokes are made of steel, while the rims may be in chromed steel (the most economical version, although this does increase the weight) or aluminum alloy. The hub is usually made of cast-aluminum alloy.

The wheel bearings, which in most cases are ball bearings, are installed in the hub housing with a slight inter-ference. A tubular element is located between their internal rings to act as a spacer. Conventional spoked wheels can be fitted only with tube-type tires. In recent years, however, new versions of spoked wheels have been made that are able to take tubeless tires.

## Solid Wheels

High-performance road motorcycles (sportbikes, sport touring, nearly all standard bikes) are fitted with single-piece wheels in which the spokes (between three and seven) are cast as a single unit with the rim and hub. The spokes are thick enough to work both under traction and compression. In this case the material used is nearly always aluminum alloy, and the complete wheel (hub, spokes, and rim) is cast.

The manufacturing technologies used are shell casting, whether by gravity or at low pressure, and in some cases (mopeds or inexpensive motorcycles) diecasting. In order to reduce mass, many wheels made for sport models have both hollow hubs and hollow spokes. Competition-motorcycle wheels differ from those of motorcycles made for road use in that their wheels are typically made of forged magnesium alloys (these have a lower density than aluminum). Recently, the wheels made for some limited-production motorcycles have been made using forged aluminum alloy.

Motorcycle wheels can also be stamped, in which case the separate parts are fastened together.

Among the most exotic, light, and expensive wheels are those made of composite materials based on carbon fiber.

## Tires

Tires have the function of supporting the load of the motorcycle and of transmitting both longitudinal and transversal forces to the ground. Furthermore, since tires are elastic elements, they also collaborate with the suspension system, making a fundamental contribution to rider

## Different Structures

In its simplest construction, a motorcycle tire has a structure composed of several superimposed plies, with the cords crossing (Pirelli).

This is the structure of a more evolved cross-ply tire, with two circumferential "belts" that increase stability in the tread area.

In radial tires the carcass is often composed of a single-ply, topped by one or two belts. A structure of this type has clear advantages over those with cross-plies.

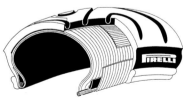

During recent years, excellent tires have appeared with a carcass composed of a ply with radial cords topped by a belt with cords at 0 degrees, meaning arranged perpendicular to the axis of rotation.

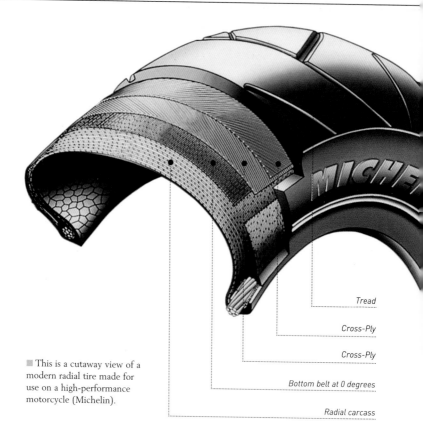

Tread

Cross-Ply

Cross-Ply

Bottom belt at 0 degrees

Radial carcass

■ This is a cutaway view of a modern radial tire made for use on a high-performance motorcycle (Michelin).

comfort. Forces are transmitted through the area of the tire that is in contact with the ground when the motorcycle is in operation (the "contact patch"); the longitudinal forces make it possible for the motorcycle to accelerate and slow down, while transversal forces act in curves. Under load (and that load changes both during acceleration and braking because of the different division of the weight on the two axles with the related raising and lowering of the front and rear suspensions), each tire deforms to a greater or lesser measure in the region of the contact patch. This means that the contact patch undergoes variations in both size and shape. The force that can be transmitted to the ground varies according to

the characteristic of the tire (the grip of the compound used for the tread), the size of the contact patch, and the load bearing on the tire itself.

*Tire Structure*
Traditionally, motorcycles were fitted with spoked wheels carrying tires with inner tubes. With the arrival of cast wheels in the 1970s, tubeless tires became commonplace, and today they dominate the scene, used on almost all road motorcycles. Tubeless tires offer the advantage of not quickly deflating when punctured but instead deflate gradually, thus making them considerably safer.

A motorcycle tire's main structure serves to bear the machine's weight and is called the carcass. The area in contact with the ground is the tread. There is a "bead" on each side of the carcass's inner diameter which seats the tire to the rim. In the case of a tubeless tire, the bead provides an airtight seal. The area between the bead and the tread is the tire's sidewall.

The carcass is composed of one or more plies, each of which is formed of a series of spirally wound threads or filaments called cords. The cords are arranged in parallel and united by an elastomer that serves as a kind of "glue." The plies are buried in an elastic matrix (which is the reason that only "rubber" is visible when one looks at the sidewalls of a tire, since the materials have a protective and sealing function). In addition to the elastomer, the elastic matrix also includes at least one pair of synthetic rubbers of different types (with less than 10 percent natural rubber), a large quantity of a thickener (lampblack), plus small quantities of various additives.

*Radials and Cross Plies*

The carcass plies can be arranged in two different ways. The cords that constitute them can be placed in a crisscross arrangement, or they can be arranged perpendicular to the centerline of the wheel. In the latter case only one or two plies are used, and the structure of the tire is called radial. One or more "belts" can be inserted above the carcass and immediately below the tread. These are circumferential strips made of a textile material that reinforce and stabilize the carcass (radial tires always have these belts).

Compared to the "classical" solution of crossed plies, tires with radial plies present significant advantages. They make it possible to have a larger contact patch with superior pressure distribution. Radial tires are also subject to less deformation; they heat up less, and they offer less rolling resistance. Additionally, radials weigh less, and provide better handling.

■ This cutaway view shows the arrangement of cords in the carcass ply with three belts (Michelin).

## Structures and Outlines

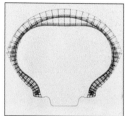

Radial tires have been given progressively lower profiles, a result of being designed with less sidewall to reduce sidewall flex.

### Tire Compounds

The qualities of the compounds composing a tire's tread (which is responsible for the transmission of forces to the ground and the tire's grip) are of great importance. The tire compound also defines the tire's optimal working temperature and its life span. Motorcycle manufacturers usually treat the formulas for tire compounds as a carefully guarded secret, especially in the case of racing tires.

A motorcycle tire's tread pattern serves several important functions, including pushing aside water from the road surface. Depending on the grooves in the tread, tires also affect traction. Tires with knob patterns are made for off-road motorcycles, and different patterns improve "grip" on different surfaces (sand, grass, gravel, and so on).

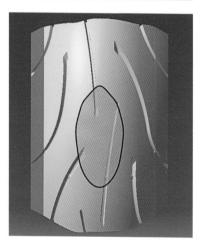

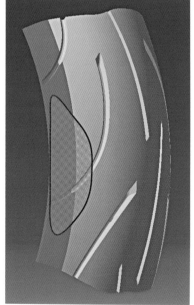

■ The contact patch is the area of tire tread in contact with the road through which transversal and longitudinal forces are transmitted. The two images show how the patch moves and changes shape when going from straight riding into a curve (Michelin).

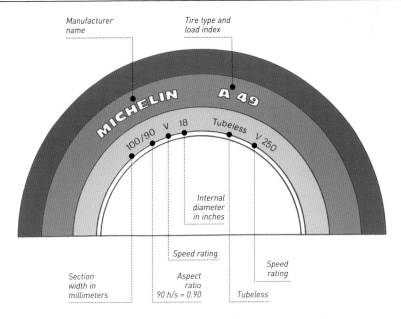

Manufacturer name

Tire type and load index

MICHELIN   A 49

100/90   V   18   Tubeless   V 250

Internal diameter in inches

Speed rating

Section width in millimeters

Aspect ratio
90 h/s = 0.90

Speed rating

Tubeless

■ This image presents the main codes used on tire sidewalls and their meaning (Michelin).

*Tire Sizes*

Tire size is indicated on the sidewall of each tire. A variety of sizing codes exist, but the metric scale is most common. In most instances, the first three digits give the section width, expressed in millimeters, followed by and separated from it by means of a back slash, the tire's aspect ratio. The aspect ratio is the percentage of the section height to the section width. Thus, a tire marked 180/60 has a nominal width of 180 millimeters and a section height of 108 millimeters (60 percent of 180). These two numbers are followed by a letter that indicates the tire's speed rating. The speed rating is usually followed by a number indicating the wheel diameter (rim size) in inches. Other codes applied to the tire sidewall indicate whether the tire is tubeless or a tube type and whether the structure is radial or cross-ply. Tires with an aspect ratio of less than 80 are called low profile.

225

# ELECTRICITY AND ELECTRONICS

Motorcycles come equipped with a variety of electrical devices. These range from the headlights to the cooling system fan, from the ignition to the dashboard instruments. To these can be added the turn indicators, the horn, and the starter motor. The ignition and injection systems are electronically controlled, as is the ABS (when present) and various anti-hopping systems.

Some touring motorcycles are equipped with electrically heated handlebar grips for use in cold weather. A recent innovation are throttle systems that control the throttle valves via a servomotor controlled by the engine's central processing unit as opposed to the traditional cable control.

All of these electronic devices make it clear that a modern motorcycle must have a strong

■ Electronic systems make it possible to run ignition and fuel-injection. Thanks to electronics some motorcycles have ABS brakes, traction-control systems, and command of the throttle valves all electronically controlled. The heart of the layout is the central processing unit (BMW).

alternator as well as a battery capable of storing electricity for those times when the engine is not operating.

## Energy Transformation

The electrical system is fed and the battery is charged by a generator that functions by converting the mechanical energy created by the engine into electricity.

Naturally, this involves a loss of power and an increase in weight and bulk. Consequently, competition engines often eliminate the generator, at least where doing so is permitted by the regulations.

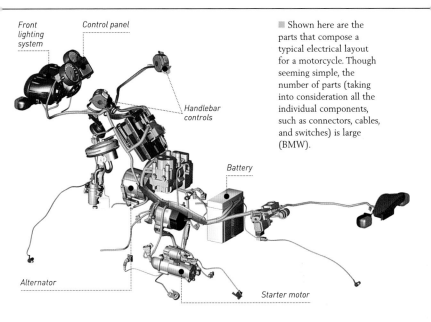

*Front lighting system*

*Control panel*

*Handlebar controls*

■ Shown here are the parts that compose a typical electrical layout for a motorcycle. Though seeming simple, the number of parts (taking into consideration all the individual components, such as connectors, cables, and switches) is large (BMW).

*Battery*

*Alternator*

*Starter motor*

Dynamos were an early electrical generating solution. These generated continuous current and made use of parts subject to wear (brushes, collectors). These were later replaced by alternators (which, as indicated by the name, produce alternating current), which are more efficient and more compact. These follow a variety of designs. Most used today are free of parts subject to wear. An alternator is composed of a rotor and a stator. The rotor is a flywheel usually splined directly to the end of the crankshaft; the stator can be connected to the side cover or fixed to the crankcase. The stator is composed of a series of copper wire coils fixed to a support plate inside which current is gen-erated. The principle behind its operation is simple and is based on the phenomenon of electromagnetic induction. A rotating element that cuts the lines of force of a magnetic field generates induced current in the copper wire coils. The induced current can be generated by magnets that turn around a series of coils (or inside it). According to the type, permanent

## Electrical Energy

An electric current, composed of electrons flowing inside a wire, is generated by a potential difference. This is called voltage and is expressed in volts. An electrical current can be used in three different ways. The first is the generation of heat, which is what happens in light bulbs when the filament is made to incandesce and thus emit light. There is also a "chemical" function: the current can be used to elicit reactions, such as that employed by batteries. The third use is by far the most important: the passage of current can be used to generate a magnetic field. Based on this application, electric motors and fuel injectors are made operable.

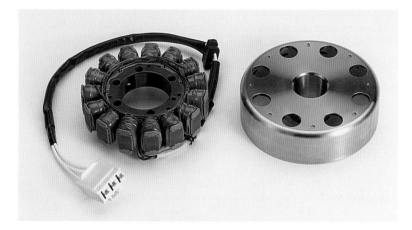

■ The alternator, which furnishes energy to the electrical circuit, has a rotor that is usually fixed to one end of the crankshaft (Kawasaki).

magnets can be used or the generator can use a coil through which current has passed, generating a magnetic field. The rotor can be internal or external. In the case of external rotors, their large size, located a good distance from the axis of rotation, lets them also perform the function of an external flywheel, fixed to one end of the crankshaft.

In some inline, four-cylinder engines, the bulk of the generator is contained by moving it from the end of the crankshaft to the rear of the crankcase, immediately behind the cylinder block. This arrangement

■ This image shows the internal structure of a battery of the type used in most motorcycles.

has also been adopted in BMW boxer twins (and on some recent Moto Guzzi bikes), with the generator operated by a poly V-belt.

Alternators generate alternating current, but batteries can store only direct current. Thus before reaching the battery the current must pass through a rectifier, which transforms alternating current into direct current. The rectifier is composed of a series of diodes (elements that allow the passage of current in one direction only). In this way the circuit and the battery are furnished with direct current only. Another important function performed by the rectifier/regulator is control of the charging system to ensure the current does not exceed predetermined limits.

### The Battery

Storing electrical energy in a place from which it can be drawn as needed is not as simple as one might imagine. The way it is done in motorcycles involves the use of batteries that function thanks to reversible chemical reactions. The arrival of current makes the reactions happen in one direction, and they are made to take place in the opposite direction when the battery is called on to provide energy. The principle has been known since the seventeenth century, but modern batteries are highly evolved devices with a great deal of applied technology.

A typical battery is an electrochemical device composed of a container divided into several compartments called cells. Inside each cell is housed a number of plates, alternately positive and negative, immersed in an electrolytic solution. The positive plates are connected by a bridge, as are the negative plates. Each plate is formed of a metal-alloy grid to which active material is applied. When the battery is charged, the grid is composed of spongy lead for the

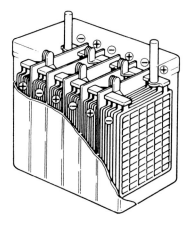

■ Every element of the battery is composed of a series of positive and negative plates that alternate inside a space (the cell) containing an electrolytic solution (Toyota).

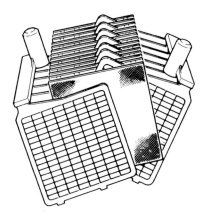

■ The plates are composed of "grids" in a conductive material on which is applied the active substance. Between one plate and another is a permeable separator (Toyota).

negative plates and lead oxide for those positive.

The output of current by the battery causes the transformation of these two substances into lead sulfate (at the same time the electrolytic solution loses sulfuric acid); when current is supplied to the battery these subjects convert again into spongy lead and lead oxide.

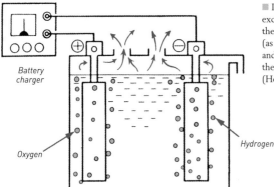

■ If a battery receives excessive charging current the plates give off gas (as shown in the figure), and the temperature of the electrolytes increases (Honda).

Battery charger

Oxygen

Hydrogen

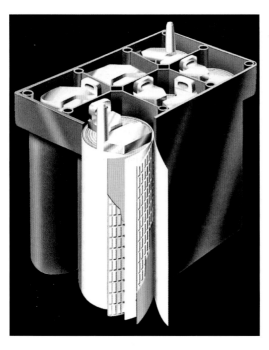

■ Recent battery designs feature cylindrical cells and "warped" plates that do not require maintenance and have an high output (Exide).

The positive and negative plates alternate and are kept apart by thin separators made of glass fiber, PVC, or packets of porous polyethylene. The electrolytic solution is composed of sulfuric acid and distilled water. Its density varies with temperature and on the basis of the relationship between the quantities of the two constituents; density increases the more the battery is charged. Every element supplies (and stores) current with a level of about 2.1 volts. A 6-volt battery is thus composed of three elements and a 12-volt of six.

An important quality of each battery is its capacity, which is expressed in amperes hour (Ah) and indicates the "amount" of energy it can store and provide (the two measurements are not

■ With the exception of race bikes, motorcycles are fitted with efficient and compact starter motors (Kawasaki).

exactly the same). The capacity, however, can vary to a certain degree on the basis of the manner by which the current is drawn. For this reason, reference is usually made to standardized conditions. The maximum absorption of current occurs when the starter motor is used. Thus a battery's "take-off" current, the amount it is able to deliver at that moment, is of notable importance.

A traditional battery is subject to self-discharge. In other words, when the motorcycle is not being used, it loses a small amount of electric energy every day. For this reason, if a motorcycle is stored for a long time, the battery should be periodically recharged. In the same way, when a motorcycle is being put to daily use, one must from time to time check to make certain the level of electrolytic solution has not decreased excessively (this loss takes place, over time, through the breather hole). Distilled water can be added to make up for such losses.

In addition to the batteries like the type just described, recent years have seen the appearance of more advanced and costly types (they function on the same principle, however). Some of these new batteries require only limited maintenance (occasional checks of the electrolytic level). Others have no need of periodic maintenance and yet others are sealed. In some instances use is made of electrolytes in the state of a gel (gel batteries). A major advantage of these is that the battery can be mounted in any position, even upside down.

■ In most cases the starter motor is connected to the crankshaft via gears and a freewheel (Suzuki).

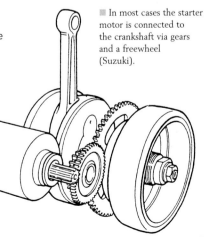

■ Motorcycle electrical systems are all 12-volt and power sophisticated lighting systems (Kawasaki).

■ Lambda sensors provide the CPU with information concerning the air/fuel mixture (KTM).

### Devices Using Electricity

Road motorcycles have been equipped with electric starters since the 1960s. Indeed, in most cases it is no longer even possible to install a kick starter. The structure of the starter is similar to that of a dynamo, although it works in the opposite way. In a starter there is a rotor, with its coil, and a stator. Starter motors are usually located on the back of the crankcase, so size and weight are restricted. For this reason they are sometimes made to turn very quickly, and the rotor is then joined to the output shaft by means of a group of epicycloidal gears. In most cases the connection to the crankshaft is entrusted to a series of gears, though sometimes a length of chain is used instead, and to a free wheel.

Other important devices using electricity are the motor of the electric cooling fan, the fuel pump, and the fuel injectors.

### Electronics

Modern motorcycles make much use of low-voltage electronic devices such as those operating ignition and injection systems. Systems like these make use of an engine central processing unit able to quickly assess an enormous quantity of information fed to it by a series of sensors. The unit's stored memory includes three-dimensional maps that give the optimum levels of ignition advance and fuel delivery (in production models these maps are more concerned with emissions control than high performance). The stored

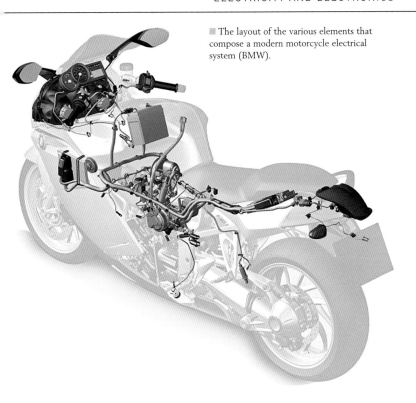

■ The layout of the various elements that compose a modern motorcycle electrical system (BMW).

information applies to the most varied conditions of use. The principal parameters are engine rpm and load, but the unit can also make adjustments on the basis of information concerning engine temperature, the air pressure in the airbox, the coolant temperature, throttle setting, and so on. Some motorcycles are equipped with two or three mappings that relate to different levels of output or response to throttle application, and these can be selected by the rider to optimize control of the motorcycle on the basis of the road surface, weather conditions, and personal taste. Electronics are also involved in the systems that control traction and antilock brakes (ABS).

# Glossary

**Alkylation**

Refinery procedure by which certain hydrocarbons (light olefins) are transformed into others of a different type (isoparaffins), whose molecules in general have from seven to nine atoms of carbon. The latter can be used to produce gasoline with excellent qualities.

**Bead seat**

The part of the wheel rim into which a tire's beads seat. Bead seats on motorcycle wheels come in standardized sizes and shapes. A tire must be mounted on a rim that has the right diameter and a bead seat of the right width.

**Butterfly valve**

In fuel-injection engines, the throttle controls the amount of air intake and does so by way of a butterfly valve. The throttle is located in a throttle body made of light alloy crossed by the intake port and connected to the head, usually by means of a rubber collar.

**Calibrated pin**

Such pins are used to position parts, as a reference or to control movement from side to side. Calibrated pins are made with great precision and absolutely minimal tolerances.

**Carbon**

Chemical element of extraordinary importance in the world of mechanics. Carbon is an essential constituent of steels and irons and even of the hydrocarbons that compose gasoline. Carbon is found in nature in the form of both diamonds (because of their extreme hardness diamonds have industrial uses, including as abrasives) and as graphite. Graphite, which is found in the cast iron used in motorcycles, is widely used in electrics (for example, in starter motor brushes) and as a solid lubricant. During recent years microcrystalline carbon has been used as the base for various kinds of coatings referred to as diamond-like carbon (DLC). There is great interest in the use of carbon fibers to make composite materials (fiber and matrix) for the manufacture of components emphasizing the traits of strength and light weight. Because of the high cost, use of carbon composites is limited to racing bikes and to some expensive production machines.

**Chrome**

A metal of great importance in the motorcycle world since it is used as an alloy in many of the better steels and in almost all superalloys. It is also widely used as a surface coating. Industrial chrome plating (not to be confused with the aesthetic kind) involves the formation of a layer of chrome with a thickness that can run from 3 to 50 microns and with a high level of hardness. Chromizing involves the application of chrome to the superficial layer of steel components. Chrome can be applied to the surface of steel components to increase hardness. Chrome passivation involves the formation of a protective film on a surface. This is done chemically by means of immersion in a bath that results in the formation of various chrome composites.

**Configuration**

The overall assembly of the components that constitute a given mechanical device. The term is also applied to a group of components, for example, the component parts of the transmission system.

**Convection**

Heat can be transmitted by conduction (through matter without any displacement of particles), by convection, or by radiation (between bodies not in contact). With convection, heat is transferred by means of a moving fluid. If the movement is obtained by means of a fan, a pump, or by the movement of the body in contact with the fluid, it is called forced convection. Heat is transferred from the external walls of motorcycle engine heads and cylinders by means of convection.

### Cracking

Process of transformation used to increase yield, meaning the quantity of gasoline obtained from a given mass of raw petroleum. Cracking can be thermal (rare today) or fluid catalytic.

### Crossflow scavenging

In two-stroke engines, crossflow scavenging takes place when the fresh air/fuel mixture enters from one side of the cylinder (through one or more ports), and the burnt gases exit from the opposite site. In this arrangement, the piston must have a deflector on its crown. The results of this scavenging are poor, and this system has been out of favor for some time. See Loop scavenging.

### Cyclic dispersion

The phases of combustion that follow one another inside the cylinder are not all equal. This is true even if the rotation speed of the engine remains the same and the throttle remains in the same position. Cycle after cycle, there are slight, but not negligible, variations in the development of heat and pressure in response to the rotations of the crankshaft. These variations constitute "cyclic dispersion" and lead to a loss of power output. Diminishing cyclic dispersion is the goal of many engineers and designers.

### Embrittlement

Following some treatments, such as tempering (temper embrittlement), the hardness of a mechanical part can be increased but at the cost of an increase in its brittleness. This is called embrittlement and sometimes it involves only the surface areas of the material.

### Engine test stand

As defined here, a laboratory facility used to develop and test engines, measuring their power and torque output. "Traditional" test stands set the engine at full throttle and alter engine rotation speeds by varying the load on the engine via the stand. There is also the inertia dynamometer, on which the complete engine is mounted with the drive wheel resting on a roller. This test reveals the performance of the engine on the basis of the acceleration given by the wheel. These are also called acceleration test stands.

### Foundry returns

Term for the extra material that must be removed to bring a part to its final size and give it the correct shape and proper surface finishing. Rough pieces from any process (casting, forging) always have a certain amount of extra material, and since its source is known it can be returned to the foundry.

### Isomerization

Refinery process by which straight-chain molecules of hydrocarbons are transformed into branched-chain molecules, yielding hydrocarbons with higher octane numbers.

### Lambda

Greek letter used to indicate the coefficient of excess air, established by the relationship between the quantity of air actually present in a determined amount of air/fuel mixture supplied to an engine and the amount that in theory should be there. If the air/fuel mixture has a stoichiometric blend, the lambda will be one. If the mixture is lean, the lambda is above one (there is more air than necessary to obtain complete combustion). In rich mixtures, the lambda is less than one.

## Loop scavenging

The system of scavenging used on modern two-stroke engines. Fresh gases enter from the side, make a large loop to the combustion chamber, flow downward, and end their route driving burnt gases out the exhaust port. The piston has a flat crown, and there are usually five transfer ports. The results are clearly superior to what is obtained with crossflow scavenging. See Crossflow scavenging.

## Offset

This is the distance between the plane on which the axes of the two fork tubes lie and the steering axis. It has an important influence on the trail, which grows as it diminishes (but the angle of inclination of the steering axis remains unchanged). It is determined by the fork triple clamps, which as a consequence can be more or less flat.

## O-ring

Sealing element made of synthetic rubber. Most typically, it is round with a round section. O-rings find wide application in motorcycle engines, providing seals to joints, small covers, and so on.

## Phosphating

Treatment that results in the formation of a layer of metallic phosphates on the surface of a mechanical component. Such a coating serves protective functions and also improves motor oil retention. The treatment is often applied to the lobes on camshafts.

## Piston rings

Elastic rings that are inserted in grooves in the piston to prevent the leakage of gases or oil between the piston and the cylinder liner.

## Polar molecules

Polar molecules have a non-uniform distribution of electric charge. As a result, one end tends to be drawn to metallic surfaces by means of electrostatic attraction.

## Press fitting

Operation in which one component is forcibly pressed inside another. The two pieces are thus interference fitted.

## Reforming

Procedure to modify the structure of hydrocarbon molecules. It is done in the presence of catalyzers.

## Seizing

Lack of lubrication or insufficient diametric clearance (the result of an error in machining or mounting or overheating) can effect piston movement to the point that there is damage to the work surfaces or that the piston even becomes locked in the cylinder. Mild seizing results in scratches or other damage from forced movement, but in severe cases there can be localized instances of melting, leading to tearing of the piston material or the inside wall of the cylinder. Seizing can also involve other areas of the engine, such as valves (and their guides) and main bearings, which can weld to the crankshaft.

## Shot blasting

Shot blasting involves the bombardment of a mechanical part with shot (made of steel, glass, or some other material) of a specified diameter. The shot is propelled by a stream of compressed air. A shot-blasted surface has a tough layer of compacted material. The most important result of shot blasting is that it increases resistance to fatigue. Motorcycle parts that are routinely shot blasted include valve springs and connecting rods.

## Silent chain

Most motorcycle chains are composed of side plates joined by rivets and sleeves; silent chain is composed of a series of parallel links joined by steel rivets. The links have teeth that engage the corresponding sprockets. Silent chains require considerable lubrication but offer a high level of reliability. As a result of the wear that takes place at the linkages after many kilometers of use, the teeth on the links engage in a more outward position on the sprockets but even so maintain the correct geometry and engage gradually, meaning they maintain their silent function. Hy-Vo and Morse are two kinds of chains in this category. Their use is typical as timing chains in single and twin overhead camshafts, but there are also examples of primary transmissions using silent chain.

## Spacer

A spacer is a tubular cylinder used to maintain the right distance between mechanical components, such as the inner rings of two bearings mounted on the same axle.

## Steady-flow rig

This experimental rig is used to test an engine head to reveal the volumes flowed by various valve lifts at different levels of pressure. The steady-flow rig is used during the design stage to optimize the fluid dynamics of intake systems.

## Stud bolt

Bolts that are screwed into a fixed component. They are used on engine heads and cylinders.

## Superalloys

Metal materials characterized by qualities that are clearly superior to those of normal steels in terms of resistance to high temperature or mechanical qualities. Since they are subject to extreme levels of heat stress, exhaust valves are often made of nickel-based superalloys. Since connecting rods are subject to extreme levels of mechanical stress, the bolts used to assemble them are often made using superalloys that have high resistance to traction (even on the order of 2000 N/mm2).

## Surface matching

One method used to ensure the cap of a connecting rod is accurately joined to the rod involves surface matching. Fractures are made at two predetermined sites on each piece thus ensuring they align when matched together.

## Timing

Operation undertaken during engine assembly to ensure that the camshaft is given the correct angular position in relation to the crankshaft. If the valve-train has not been timed properly, the valves will not open and close as called for in the original engine design.

## Toothed belt

Toothed belts are used in motorcycle transmissions, as timing belts, and as final-drive. Unlike situations employing trapezoidal belts, toothed belts do not require pulleys. A typical toothed belt is composed of a series of resistant inserts, called cords, embedded in a body of synthetic rubber so that the inner surface of the belt will present teeth of a carefully designed size and shape. The surface of the teeth is protected from wear by one or more layers of nylon or some other material. Toothed belts are lightweight, inexpensive, and they work silently. They do not require lubrication, and in theory, they cannot be stretched. The belts are wide and must not come in contact with gasoline, motor oil, or other petroleum-based substances. They require periodic replacement.

## Traction

Also known as grip, this is the capacity of a tire to transmit forces to the ground. It is closely related to the qualities of the compound used in the manufacture of the tread and increases with increases in the load bearing down on the tire itself. A tire's traction is highly influenced by road conditions. The forces in question are both longitudinal (that is braking and accelerating) and lateral.

## Triple bridge

Brake calipers with four opposing pistons in which the two main parts (the two calipers) are joined not just at the ends but also by a central bridge. This configuration increases the structural rigidity of the component.

**Valve lash**
When a valve is closed, there must be a small gap, or clearance, between the rocker arm (or tappet) and the valve stem. This gap ensures that the valve always returns to its seat. Because of thermal dilation, the parts of the valve-train change shape and consequently the lash can vary, but it must never be eliminated. If that were to occur, the valve

would no longer provide a leak-free seal, resulting in a decline in performance as well as overheating, with the risk of damage (a "burned" valve).

**Variable rate spring**
When the distance between consecutive coils of a spring is not constant but varies for the length of the spring, it is said to have a variable rate.

**Wrist pin**
Tubular steel pin that joins the piston to the connecting rod.